JEFFERY A. COLE

PARTIAL SOLUTIONS MANUAL

to accompany

PRECALCULUS
FUNCTIONS AND GRAPHS

FIFTH EDITION

E A R L   W.   S W O K O W S K I

 PWS-KENT PUBLISHING COMPANY
Boston

PWS-KENT
Publishing Company

20 Park Plaza
Boston, Massachusetts 02116

PWS-KENT Publishing Company is a division of Wadsworth, Inc.

ISBN 0-87150-147-3

Printed in the United States of America

88 89 90 91 -- 10 9 8 7 6 5 4 3

Preface

This manual contains solutions to approximately one-third of the exercises in the text, *Precalculus : Functions and Graphs*, Fifth Edition, by Earl W. Swokowski. The general pattern of problem solutions included is 1, 4, 7, ..., $3n + 1$, ..., where n is a nonnegative integer. This pattern is sometimes supplemented by additional solutions or changed so that duplicate solutions are avoided.

A significant number of today's students are involved in various outside activities, and find it difficult, if not impossible, to attend all class sessions. This manual should help meet the needs of these students. In addition, it is my hope that this manual's solutions will enhance the understanding of all readers of the material and provide insights to solving other exercises.

A great amount of effort has gone into the accuracy of the solutions. I would appreciate any feedback concerning errors, solution correctness, or solution style. These and any other comments may be sent to me at the address below or in care of the publisher.

I would like to thank the PWS–KENT editorial staff and Earl Swokowski for entrusting me with this project. In particular, I would like to thank Dee Hart for handling all the editorial tasks. I dedicate this book to my loving parents, Marlin and Marilyn, and thank them for their continued moral support and understanding.

Jeffery A. Cole

Anoka-Ramsey Community College

11200 Mississippi Blvd. NW

Coon Rapids, MN 55433

Table of Contents

1) (a) $-7 < -4$ (b) $3 > -1$ (c) $1 + 3 = 6 - 2$ since $4 = 4$

4) (a) $\frac{1}{7} < 0.143$ since $\frac{1}{7} \simeq 0.142857$ {$\simeq$ means "approximately equal to"}

(b) $\frac{3}{4} + \frac{2}{3} < \frac{18}{12}$ since $\frac{3}{4} + \frac{2}{3} = \frac{9}{12} + \frac{8}{12} = \frac{17}{12}$

(c) $\sqrt{2} > 1.4$ since $\sqrt{2} \simeq 1.414214$

7) 0 is greater than -1 can be written as $0 > -1$

10) y is nonnegative can be written as $y \geq 0$

13) $(-2, 1) \iff -2 < x < 1$ {$\iff$ means "is equivalent to"}

16) $[-3, -2] \iff -3 \leq x \leq -2$

19) $(-\infty, 2) \iff x < 2$

22) $(-\infty, 1] \iff x \leq 1$

25) $4 \geq x \geq 1 \iff [1, 4]$ 28) $x \leq 2 \iff (-\infty, 2]$

Figure 25   Figure 28

31) (a) $|4 - 9| = -(4 - 9)$ {since $4 - 9 < 0$} $= -(-5) = 5$

(b) $|-4| - |-9| = -(-4) - (-(-9)) = 4 - 9 = -5$

(c) $|4| + |-9| = 4 + -(-9) = 4 + 9 = 13$

34) (a) $|8 - 5| = 8 - 5$ {since $8 - 5 \geq 0$} $= 3$

(b) $-5 + |-7| = -5 + -(-7) = -5 + 7 = 2$

(c) $(-2) |-2| = (-2) (-(-2)) = (-2) (2) = -4$

For Exercises 37, 40, and 43 use the Quadratic Formula

$x = \dfrac{-b \pm \sqrt{b^2 - 4ac}}{2a}$ to find the solutions to $ax^2 + bx + c = 0$.

37) $2x^2 - x - 3 = 0$ $(a = 2, b = -1, c = -3)$

$x = \dfrac{1 \pm \sqrt{1 + 24}}{4} = \dfrac{1 \pm 5}{4} = \dfrac{6}{4}, \dfrac{-4}{4} = \dfrac{3}{2}, -1$

40) $v^2 + 3v - 5 = 0$ $(a = 1, b = 3, c = -5)$

Exercises 1.1

$$v = \frac{-3 \pm \sqrt{9 + 20}}{2} = \frac{-3 \pm \sqrt{29}}{2}$$

43) $4y^2 - 20y + 25 = 0$ $(a = 4, b = -20, c = 25)$

$$y = \frac{20 \pm \sqrt{400 - 400}}{8} = \frac{20 \pm 0}{8} = \frac{20}{8} = \frac{5}{2}$$

46) Let l be the length of the side parallel to the river bank and let w be the length of each of the sides adjacent to l.

   (a) $P = 2w + l = 2w + 2w = 4w$ {formula for the perimeter}

   $4w = 180 \implies w = 45$ ft. {$\implies$ means "implies"}

   $A = (45)(90) = 4050$ sq. ft. {Area = width X length}

   (b) $P = 2w + l = 2w + \frac{1}{2}w = \frac{5}{2}w$; $\frac{5}{2}w = 180 \implies w = 72$ ft.

   $A = (72)(36) = 2592$ sq. ft.

   (c) $P = 2w + l = 2w + w = 3w$; $3w = 180 \implies w = 60$ ft.

   $A = (60)(60) = 3600$ sq. ft.

49) (a) let $v = 55$; $d = 55 + \frac{55^2}{20} = 206.25$ ft.

   (b) let $d = 120$; $120 = v + \frac{v^2}{20} \implies v^2 + 20v - 2400 = 0 \implies$

$$v = \frac{-20 \pm \sqrt{400 + 9600}}{2} = \frac{-20 \pm 100}{2} = \frac{80}{2}, \frac{-120}{2} = 40, -60;$$

   Both numbers are solutions to the equation, but the solution of

   40 mph is the only one that makes sense for this problem.

Exercises 1.2

1) $(3u^7v^3)(4u^4v^{-5}) = 12u^{7+4}v^{3+(-5)} = 12u^{11}v^{-2} = \frac{12u^{11}}{v^2}$

4) $\left(\frac{4a^2b}{a^3b^2}\right)\left(\frac{5a^2b}{2b^4}\right) = \left(\frac{20a^4b^2}{2a^3b^6}\right) = \frac{10a}{b^4}$

7) $\left(\dfrac{3x^5y^4}{x^0y^{-3}}\right)^2 = \left(\dfrac{3^2x^{10}y^8}{x^0y^{-6}}\right) = 9x^{10}y^{14}$

10) $(25z^4)^{-3/2} = \dfrac{1}{(25z^4)^{3/2}} = \dfrac{1}{(\sqrt{25z^4})^3} = \dfrac{1}{(5z^2)^3} = \dfrac{1}{125z^6}$

13) $\sqrt[3]{8a^6b^{-3}} = (8a^6b^{-3})^{1/3} = 8^{1/3}a^{6/3}b^{-3/3} = 2a^2b^{-1} = \dfrac{2a^2}{b}$

16) $\sqrt[3]{\dfrac{1}{4x^5y^2}} = \sqrt[3]{\dfrac{1}{4x^5y^2} \cdot \dfrac{2xy}{2xy}}$ {multiply by 2xy so that the denominator contains only "cube" terms}

$= \sqrt[3]{\dfrac{2xy}{8x^6y^3}} = \dfrac{\sqrt[3]{2xy}}{2x^2y}$

19) $\sqrt[3]{(a+b)^2} = (a+b)^{2/3}$

22) (a) $4 + x^{3/2} = 4 + x^{2/2}x^{1/2} = 4 + x\sqrt{x}$

(b) $(4+x)^{3/2} = (4+x)^{2/2}(4+x)^{1/2} = (4+x)\sqrt{4+x}$

25) $a^xb^y \neq (ab)^{xy}$ since $(ab)^{xy} = a^{xy}b^{xy}$

28) $(2x^5 - 3x^2 + 2) + (-2x^4 + x^2 - 4x + 5) = (2x^5) + (-2x^4) +$

$(-3x^2 + x^2) + (-4x) + (2 + 5) = 2x^5 - 2x^4 - 2x^2 - 4x + 7;$

the degree of the last polynomial is 5

31) $(2x^3 - x + 5)(x^2 + x + 2) = 2x^3(x^2 + x + 2) + (-x)(x^2 + x + 2) +$

$5(x^2 + x + 2) = 2x^5 + 2x^4 + 4x^3 - x^3 - x^2 - 2x + 5x^2 + 5x + 10 =$

$2x^5 + 2x^4 + 3x^3 + 4x^2 + 3x + 10;$ degree is 5

34) $(x + 9y^2)(3x - 4y^2) = x(3x - 4y^2) + 9y^2(3x - 4y^2) = 3x^2 - 4xy^2 +$

$27xy^2 - 36y^4 = 3x^2 + 23xy^2 - 36y^4$

37) $(x - 2y)^3 = (x)^3 - 3(x)^2(2y)^1 + 3(x)^1(2y)^2 - (2y)^3 =$

$x^3 - 6x^2y + 12xy^2 - 8y^3$ {Using Product Formula (vi) on page 17}

40) $10x^2 + 29x - 21 = (5x - 3)(2x + 7)$

43) $25x^2 - 9 = (5x + 3)(5x - 3)$

Exercises 1.2

46) $2x^3 - x^2 - 2x + 1 = x^2(2x - 1) - 1(2x - 1) = (x^2 - 1)(2x - 1) = (x + 1)(x - 1)(2x - 1)$

49) Recognizing this as the difference of squares, we have $x^6 - 1 = (x^3)^2 - (1)^2 = (x^3 + 1)(x^3 - 1)$ { these are the sum of cubes and the difference of cubes, respectively } $=$

$(x + 1)(x^2 - x + 1)(x - 1)(x^2 + x + 1)$ ★

If this is treated as a difference of cubes first, the steps are as follows : $x^6 - 1 = (x^2)^3 - (1)^3 = (x^2 - 1)(x^4 + x^2 + 1)$ ★★

To factor $x^4 + x^2 + 1$, we add x^2 and subtract x^2 to obtain

$(x^4 + 2x^2 + 1) - x^2 = (x^2 + 1)^2 - (x)^2$ and treat this as a difference of squares; substituting into ★★ yields ★.

52) $\dfrac{10x^2 + 29x - 21}{5x^2 - 23x + 12} = \dfrac{(5x - 3)(2x + 7)}{(5x - 3)(x - 4)} = \dfrac{2x + 7}{x - 4}$

55) $\dfrac{6 - 7a - 5a^2}{10a^2 - a - 3} = \dfrac{-(5a^2 + 7a - 6)}{10a^2 - a - 3} = \dfrac{-(5a - 3)(a + 2)}{(5a - 3)(2a + 1)} = \dfrac{-(a + 2)}{2a + 1}$

58) $\dfrac{16x^4 + 8x^3 + x^2}{4x^3 + 25x^2 + 6x} = \dfrac{x^2(16x^2 + 8x + 1)}{x(4x^2 + 25x + 6)} = \dfrac{x^2(4x + 1)^2}{x(4x + 1)(x + 6)} =$

$\dfrac{x(4x + 1)}{x + 6}$

61) $\dfrac{2x + 1}{2x - 1} - \dfrac{x - 1}{x + 1} = \dfrac{(2x + 1)(x + 1) - (x - 1)(2x - 1)}{(2x - 1)(x + 1)} =$

$\dfrac{(2x^2 + 3x + 1) - (2x^2 - 3x + 1)}{(2x - 1)(x + 1)} = \dfrac{6x}{(2x - 1)(x + 1)}$

64) $\dfrac{a^2 + 4a + 3}{3a^2 + a - 2} \cdot \dfrac{3a^2 - 2a}{2a^2 + 13a + 21} = \dfrac{(a + 3)(a + 1)}{(3a - 2)(a + 1)} \cdot \dfrac{a(3a - 2)}{(2a + 7)(a + 3)} =$

$\dfrac{a}{2a + 7}$

67) $\dfrac{2}{3x + 1} - \dfrac{9}{(3x + 1)^2} = \dfrac{2(3x + 1) - 9}{(3x + 1)^2} = \dfrac{6x - 7}{(3x + 1)^2}$

70) $\dfrac{6}{3t} + \dfrac{t+5}{t^3} + \dfrac{1-2t^2}{t^4}$ { reduce the first term to $\dfrac{2}{t}$ and then use t^4 as

the least common denominator } $= \dfrac{2\,(t^3) + (t+5)\,(t) + (1-2t^2)}{t^4} =$

$\dfrac{2t^3 - t^2 + 5t + 1}{t^4}$

73) $\dfrac{2}{x} + \dfrac{7}{x^2} + \dfrac{5}{2x-3} + \dfrac{1}{(2x-3)^2} =$

$\dfrac{2x\,(2x-3)^2 + 7\,(2x-3)^2 + 5x^2\,(2x-3) + 1x^2}{x^2\,(2x-3)^2} =$

$\dfrac{2x\,(4x^2 - 12x + 9) + 7\,(4x^2 - 12x + 9) + 10x^3 - 15x^2 + x^2}{x^2\,(2x-3)^2} =$

$\dfrac{(8x^3 - 24x^2 + 18x) + (28x^2 - 84x + 63) + 10x^3 - 14x^2}{x^2\,(2x-3)^2} =$

$\dfrac{18x^3 - 10x^2 - 66x + 63}{x^2\,(2x-3)^2}$

76) $2 + \dfrac{3}{x} + \dfrac{7x}{3x+10} = \dfrac{2\,(x)\,(3x+10) + 3\,(3x+10) + 7x\,(x)}{x\,(3x+10)} =$

$\dfrac{6x^2 + 20x + 9x + 30 + 7x^2}{x\,(3x+10)} = \dfrac{13x^2 + 29x + 30}{x\,(3x+10)}$

79) $\dfrac{\dfrac{x}{y^2} - \dfrac{y}{x^2}}{\dfrac{1}{y^2} - \dfrac{1}{x^2}} = \dfrac{\dfrac{x\,(x^2) - y\,(y^2)}{x^2 y^2}}{\dfrac{1\,(x^2) - 1\,(y^2)}{x^2 y^2}} = \dfrac{\dfrac{x^3 - y^3}{x^2 y^2}}{\dfrac{x^2 - y^2}{x^2 y^2}} = \dfrac{x^3 - y^3}{x^2 y^2} \cdot \dfrac{x^2 y^2}{x^2 - y^2} =$

$\dfrac{(x-y)\,(x^2 + xy + y^2)}{(x-y)\,(x+y)} = \dfrac{x^2 + xy + y^2}{x+y}$

82) $\dfrac{\dfrac{3}{w} - \dfrac{6}{2w+1}}{\dfrac{5}{w} + \dfrac{8}{2w+1}} = \dfrac{\dfrac{3\,(2w+1) - 6\,(w)}{w\,(2w+1)}}{\dfrac{5\,(2w+1) + 8\,(w)}{w\,(2w+1)}} = \dfrac{3}{w\,(2w+1)} \cdot \dfrac{w\,(2w+1)}{18w+5} =$

$\dfrac{3}{18w+5}$

Exercises 1.2

85) $\dfrac{\dfrac{1}{(x+h)^3} - \dfrac{1}{x^3}}{h} = \dfrac{\dfrac{x^3 - (x+h)^3}{x^3(x+h)^3}}{h} = \dfrac{x^3 - (x^3 + 3x^2h + 3xh^2 + h^3)}{x^3(x+h)^3} \cdot \dfrac{1}{h} =$

$\dfrac{-3x^2h - 3xh^2 - h^3}{hx^3(x+h)^3} = \dfrac{-h(3x^2 + 3xh + h^2)}{hx^3(x+h)^3} = \dfrac{-(3x^2 + 3xh + h^2)}{x^3(x+h)^3}$

88) $(6x-5)^3 (2)(x^2+4)(2x) + (x^2+4)^2 (3)(6x-5)^2 (6) =$

$(6x-5)^2 (2)(x^2+4) [(6x-5)(2x) + (x^2+4)(3)(3)] =$

$2(6x-5)^2 (x^2+4) [12x^2 - 10x + 9x^2 + 36] =$

$2(6x-5)^2 (x^2+4)(21x^2 - 10x + 36)$

91) $\dfrac{(6x+1)^3 (27x^2+2) - (9x^3+2x)(3)(6x+1)^2 (6)}{(6x+1)^6} =$

$\dfrac{(6x+1)^2 [(6x+1)(27x^2+2) - (9x^3+2x)(3)(6)]}{(6x+1)^6} =$

$\dfrac{[(162x^3 + 27x^2 + 12x + 2) - (162x^3 + 36x)]}{(6x+1)^4} = \dfrac{27x^2 - 24x + 2}{(6x+1)^4}$

94) $\dfrac{(1-x^2)^{1/2}(2x) - x^2 (\frac{1}{2})(1-x^2)^{-1/2}(-2x)}{[(1-x^2)^{1/2}]^2} = $ { $(1-x^2)^{-1/2}(x)$ is the greatest common fac.}

$\dfrac{(1-x^2)^{-1/2}(x) [(1-x^2)(2) - (x)(-x)]}{(1-x^2)^1} = \dfrac{x(2-x^2)}{(1-x^2)^{3/2}}$

97) $\dfrac{\sqrt{a} - \sqrt{b}}{c} \cdot \dfrac{\sqrt{a} + \sqrt{b}}{\sqrt{a} + \sqrt{b}} = \dfrac{(\sqrt{a})^2 - (\sqrt{b})^2}{c(\sqrt{a} + \sqrt{b})} = \dfrac{a - b}{c(\sqrt{a} + \sqrt{b})}$

100) $\dfrac{\sqrt{1-x-h} - \sqrt{1-x}}{h} \cdot \dfrac{\sqrt{1-x-h} + \sqrt{1-x}}{\sqrt{1-x-h} + \sqrt{1-x}} =$

$\dfrac{(1-x-h) - (1-x)}{h(\sqrt{1-x-h} + \sqrt{1-x})} = \dfrac{-h}{h(\sqrt{1-x-h} + \sqrt{1-x})} =$

$\dfrac{-1}{\sqrt{1-x-h} + \sqrt{1-x}}$

1) $5x - 6 > 11 \Rightarrow 5x > 17 \Rightarrow x > \frac{17}{5}; \quad (\frac{17}{5}, \infty)$

4) $7 - 2x \geq -3 \Rightarrow -2x \geq -10 \Rightarrow x \leq 5; \quad (-\infty, 5]$

7) $-4 < 3x + 5 < 8 \Rightarrow -9 < 3x < 3 \Rightarrow -3 < x < 1; \quad (-3, 1)$

10) $-2 \leq \dfrac{5 - 3x}{4} \leq \dfrac{1}{2} \Rightarrow -8 \leq 5 - 3x \leq 2 \Rightarrow -13 \leq -3x \leq -3 \Rightarrow$

$\frac{13}{3} \geq x \geq 1 \Rightarrow 1 \leq x \leq \frac{13}{3}; \quad [1, \frac{13}{3}]$

13) $|x - 10| < 0.05 \Rightarrow -0.05 < x - 10 < 0.05 \Rightarrow$

$9.95 < x < 10.05; \quad (9.95, 10.05)$

16) $|7x + 4| < 10 \Rightarrow -10 < 7x + 4 < 10 \Rightarrow -14 < 7x < 6 \Rightarrow$

$-2 < x < \frac{6}{7}; \quad (-2, \frac{6}{7})$

19) $x^2 - x - 6 < 0 \Rightarrow (x - 3)(x + 2) < 0$

```
Sign of the product            +    |   -   |   +
Sign of the factor (x - 3)   - - -  | - - - | + + +
Sign of the factor (x + 2)   - - -  | + + + | + + +
                               -2       3
```

The product $(x - 3)(x + 2)$ is less than 0 on the interval $(-2, 3)$.

22) $4x^2 \geq x \Rightarrow 4x^2 - x \geq 0 \Rightarrow (x)(4x - 1) \geq 0$

```
Sign of the product            +    |   -   |   +
Sign of the factor (4x - 1)  - - -  | - - - | + + +
Sign of the factor (x)         - - -  | + + + | + + +
                                 0       1/4
```

The product $(x)(4x - 1)$ is greater than or equal to 0 on the

intervals $(-\infty, 0]$ and $[\frac{1}{4}, \infty)$. $\quad (-\infty, 0] \cup [\frac{1}{4}, \infty)$

25) $\dfrac{x + 4}{2x - 1} < 3 \Rightarrow \dfrac{x + 4}{2x - 1} - 3 < 0 \Rightarrow \dfrac{x + 4}{2x - 1} - \dfrac{3(2x - 1)}{2x - 1} < 0 \Rightarrow$

$\dfrac{-5x + 7}{2x - 1} < 0$

Exercises 1.3

Sign of the quotient $-$ | $+$ | $-$
Sign of the factor $(-5x + 7)$ $+ + +|+ + +|- - -$
Sign of the factor $(\ 2x - 1)$ $- - -|+ + +|+ + +$
                                      $1/2$ $7/5$

The quotient $\dfrac{-5x + 7}{2x - 1}$ is less than 0 on the intervals $(-\infty, \frac{1}{2})$ and

$(\frac{7}{5}, \infty)$; $(-\infty, \frac{1}{2}) \cup (\frac{7}{5}, \infty)$

28) $(x^2 - 4x + 4)(3x - 7) < 0 \Rightarrow (x - 2)^2 (3x - 7) < 0$; Now $(x - 2)^2$
is always greater than zero except when $x = 2$. Thus, we only need
the solution of $3x - 7 < 0$, which is $x < \frac{7}{3}$; Remembering to exclude
the value of 2, we have $(-\infty, 2) \cup (2, \frac{7}{3})$.

31) Since $C = \frac{5}{9}(F - 32)$, we want to subtract 32 and then multiply the
resulting values by $\frac{5}{9}$ to obtain the equivalent Celsius temperatures.
$60 \le F \le 80 \Rightarrow 60 - 32 \le F - 32 \le 80 - 32 \Rightarrow$
$\frac{5}{9}(28) \le \frac{5}{9}(F - 32) \le \frac{5}{9}(48) \Rightarrow \frac{140}{9} \le C \le \frac{80}{3}$

34) $\dfrac{1}{R} = \dfrac{1}{R_1} + \dfrac{1}{R_2} = \dfrac{R_2 + R_1}{R_1 R_2}$; Taking reciprocals of both sides we have

$R = \dfrac{R_1 R_2}{R_2 + R_1}$. Now let $R_1 = 10$ to obtain $R = \dfrac{10 R_2}{R_2 + 10}$. Since we

want the values of R that are less than 5, solve $R < 5$. i.e.

$\dfrac{10 R_2}{R_2 + 10} < 5 \Rightarrow \dfrac{10 R_2 - 5 (R_2 + 10)}{R_2 + 10} < 0 \Rightarrow \dfrac{5 R_2 - 50}{R_2 + 10} < 0 \Rightarrow$

$\dfrac{R_2 - 10}{R_2 + 10} < 0$; Solving this inequality as in Exercise 25 yields the

solution $-10 < R_2 < 10$. Now R_2 can only be positive, $\{R_2 = 0$
would make the original equation undefined$\}$ so the final solution is
the range of values $0 < R_2 < 10$.

37) We are given $d = v + \dfrac{v^2}{20}$ and we want to know when the distance (d) will be less than 75. Solve d < 75. $v + \dfrac{v^2}{20} < 75 \Rightarrow$

$20v + v^2 < 1500 \Rightarrow v^2 + 20v - 1500 < 0 \Rightarrow$

$(v + 50)(v - 30) < 0$; Solving this inequality yields the solution $-50 < v < 30$, but v must be nonnegative. Thus the final solution is the range of values $0 \le v < 30$.

40) We want to know when there will be more of gas B than gas A and we have 10 moles of gas A to begin with. As we "lose" gas A, we we "gain" gas B. We will have more of gas B when this amount is greater than one-half of the original amount of gas A. Solve B > 5.

$\dfrac{10t}{t + 4} > \dfrac{1}{2}(10) \Rightarrow \dfrac{10t}{t + 4} - 5 > 0 \Rightarrow \dfrac{10t - 5(t + 4)}{t + 4} > 0 \Rightarrow$

$\dfrac{5t - 20}{t + 4} > 0 \Rightarrow \dfrac{t - 4}{t + 4} > 0$; Solving this inequality, we obtain

t < -4 or t > 4. Since t is nonnegative, there will be more of gas B after 4 minutes.

1)

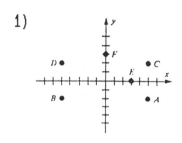

Figure 1

4) The set of points (x, -x) forms a line that bisects quadrants II and IV.

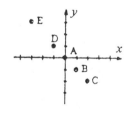

Figure 4

7) Given the points A (4, -3) and B (6, 2), we have :

(a) $d(A, B) = \sqrt{(6 - 4)^2 + (2 - (-3))^2} = \sqrt{2^2 + 5^2} + \sqrt{4 + 25} = \sqrt{29}$

Exercises 1.4

(b) midpoint $= \left(\dfrac{4+6}{2}, \dfrac{-3+2}{2} \right) = (5, -\tfrac{1}{2})$

10) Given the points A (6, 2) and B (6, –2), we have :

   (a) d (A, B) $= \sqrt{(6-6)^2 + (-2-2)^2} = \sqrt{0^2 + (-4)^2} = \sqrt{0 + 16} = 4$

   (b) midpoint $= \left(\dfrac{6+6}{2}, \dfrac{2+(-2)}{2} \right) = (6, 0)$

13) We must show that the three sides of the triangle determined by the three points satisfy the Pythagorean Theorem.

   d (A, B) $= \sqrt{(-7)^2 + (-7)^2} = \sqrt{98}$; d (B, C) $= \sqrt{(-4)^2 + 4^2} = \sqrt{32}$;

   d (A, C) $= \sqrt{(-11)^2 + (-3)^2} = \sqrt{130}$;

   Now since $(\sqrt{98})^2 + (\sqrt{32})^2 = 98 + 32 = 130$ and $(\sqrt{130})^2 = 130$, we see that a right triangle is formed. To find the area, we use the formula A $= \tfrac{1}{2}$ bh. { Area $= \tfrac{1}{2}$ X base X height } The two shorter sides represent the base and the height of the triangle, so

   A $= \tfrac{1}{2} \sqrt{98} \sqrt{32} = \tfrac{1}{2} (7\sqrt{2})(4\sqrt{2}) = 14 \cdot 2 = 28$ square units.

16) The graph is a line with x-intercept at $\tfrac{3}{4}$ and y-intercept at –3. The original equation is y = 4x – 3. Substituting –x for x yields y = –4x – 3. Substituting –y for y yields –y = 4x – 3 or equivalently, y = –4x + 3. Substituting –x for x and –y for y yields –y = –4x – 3 or equivalently, y = 4x + 3. None of these substitutions result in an equivalent equation of the original, so there is no symmetry for this graph. See Figure 16.

19) Substituting –x for x leads to an equivalent equation, so the graph is symmetric with respect to the y-axis. The y-intercept is –1 and the x-intercepts are $\pm \sqrt{2}/2$ { $\simeq 0.707$ }. See Figure 19.

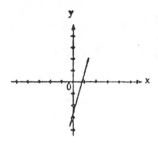

Figure 16

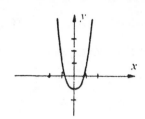

Figure 19

22) Substituting −x for x leads to an equivalent equation, so the graph is is symmetric with respect to the y-axis. The only intercept of the parabola is the origin. See Figure 22.

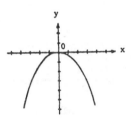

Figure 22

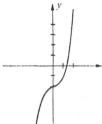

Figure 25

25) There is no symmetry in this graph. The y-intercept is −2 and the x-intercept is $\sqrt[3]{2}$ {≃ 1.26}. See Figure 25.

28) There is no symmetry in this graph. The y-intercept is −1 and the x-intercept is 1. See Figure 28.

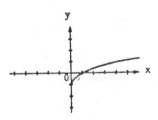

Figure 28

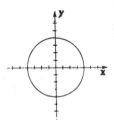

Figure 31

Exercises 1.4

31) Since $(-x)^2 = x^2$ and $(-y)^2 = y^2$, any of the symmetry tests yield an equivalent equation. Therefore, the graph is symmetric with respect to the y-axis, the x-axis, and the origin. The graph is a circle with radius 4 and center at the origin. See Figure 31.

34) Substituting $-x$ for x leads to an equivalent equation, so the graph is symmetric with respect to the y-axis. The graph is a semi-circle with x-intercepts at ± 2 and a y-intercept at -2. See Figure 34.

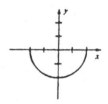

Figure 34

37) With a center at C $(3, -2)$ and a radius of 4, we have the equation of the circle as $(x - 3)^2 + (y - (-2))^2 = 4^2$ or $(x - 3)^2 + (y + 2)^2 = 16$.

40) With a center of $(-4, 6)$, we know that the equation must be of the form $(x + 4)^2 + (y - 6)^2 = r^2$. Since the circle passes through P $(1, 2)$, we know that $x = 1$ and $y = 2$ is one solution of the general equation. Substituting $x = 1$ and $y = 2$ yields : $5^2 + (-4)^2 = r^2 \Rightarrow r^2 = 41$. $(x + 4)^2 + (y - 6)^2 = 41$ is a circle with radius $\sqrt{41}$.

43) The center of the circle is the midpoint of the endpoints.

$$\text{midpoint} = \left(\frac{4 + -2}{2}, \frac{-3 + 7}{2}\right) = (1, 2);$$ The distance from A to B is $\sqrt{(-6)^2 + 10^2} = \sqrt{136} = 2\sqrt{34}$. One-half that distance is the radius, i.e. $r = \sqrt{34}$. The equation is $(x - 1)^2 + (y - 2)^2 = 34$.

46) $x^2 + y^2 - 10x + 2y + 22 = 0 \iff$ {complete the square on x and y}
$(x^2 - 10x + \underline{25}) + (y^2 + 2y + \underline{1}) = -22 + \underline{25} + \underline{1} \iff$
$(x - 5)^2 + (y + 1)^2 = 4$; A circle with C $(5, -1)$ and $r = \sqrt{4} = 2$.

49) $2x^2 + 2y^2 - x + y - 3 = 0$ ⟷ { divide by 2 }

$x^2 - \frac{1}{2}x + y^2 + \frac{1}{2}y = \frac{3}{2}$ ⟷ { complete the square on x and y }

$(x^2 - \frac{1}{2}x + \frac{1}{16}) + (y^2 + \frac{1}{2}y + \frac{1}{16}) = \frac{3}{2} + \frac{1}{16} + \frac{1}{16}$ ⟷

$(x - \frac{1}{4})^2 + (y + \frac{1}{4})^2 = \frac{26}{16}$; A circle with C $(\frac{1}{4}, -\frac{1}{4})$ and $r = \sqrt{26}/4$.

1) $m = \dfrac{18 - 6}{(-1) - (-4)} = \dfrac{12}{3} = 4$ $\qquad$ 4) $m = \dfrac{4 - 4}{2 - (-3)} = \dfrac{0}{5} = 0$

7) The slopes of opposite sides are equal (parallel) and the slopes of
of two adjacent sides are negative reciprocals (perpendicular).

$m_{AB} = m_{CD} = -\frac{3}{5}$ and $m_{BC} = m_{AD} = \frac{5}{3}$

10) Let $E = M (A, B) = \left[\dfrac{x_1 + x_2}{2}, \dfrac{y_1 + y_2}{2}\right]$, { where M (A, B) stands for
$\qquad\qquad\qquad\qquad\qquad\qquad\qquad\qquad$ the midpoint of A and B }

$\qquad F = M (B, C) = \left[\dfrac{x_2 + x_3}{2}, \dfrac{y_2 + y_3}{2}\right]$,

$\qquad G = M (C, D) = \left[\dfrac{x_3 + x_4}{2}, \dfrac{y_3 + y_4}{2}\right]$, and

$\qquad H = M (C, D) = \left[\dfrac{x_1 + x_4}{2}, \dfrac{y_1 + y_4}{2}\right]$.

m_{EF} {the slope between E and F} $= \dfrac{\frac{y_2 + y_3}{2} - \frac{y_1 + y_2}{2}}{\frac{x_2 + x_3}{2} - \frac{x_1 + x_2}{2}} =$

$\dfrac{\frac{y_3 - y_1}{2}}{\frac{x_3 - x_1}{2}} = \dfrac{y_3 - y_1}{x_3 - x_1}$; Similarly, m_{GH}, m_{FG}, and m_{EH} can be found.

$m_{EF} = m_{GH} = \dfrac{y_3 - y_1}{x_3 - x_1}$ and $m_{FG} = m_{EH} = \dfrac{y_4 - y_2}{x_4 - x_2}$

The slopes of opposite sides are equal, so we have a parallelogram.

13) $m_{AB} = \frac{3}{8}$; Using the Point-Slope Formula with A $(-5, -7)$, we have

$y - (-7) = \frac{3}{8}(x - (-5))$ ⟺ $8(y + 7) = 3(x + 5)$ ⟺
$3x - 8y - 41 = 0$.

16) An x-intercept of -2 is equivalent to having the point $(-2, 0)$ on the
line. Applying the Point-Slope Formula yields $y - 0 = 6(x - (-2))$
⟺ $y = 6x + 12$ ⟺ $6x - y + 12 = 0$.

19) Solving $2x - 5y = 8$ for y yields the equation $y = \frac{2}{5}x - \frac{8}{5}$. The slope
of this line is $\frac{2}{5}$, so the perpendicular slope is $-\frac{5}{2}$. Applying the
Point-Slope Formula yields $y - (-3) = -\frac{5}{2}(x - 7)$ ⟺
$2(y + 3) = -5(x - 7)$ ⟺ $5x + 2y - 29 = 0$.

22) Recognizing that this line must have a slope of -1 and pass through
the origin, we have $y - 0 = -1(x - 0)$ ⟺ $y = -x$ ⟺ $x + y = 0$.

25) $x + 2y = 0$ ⟺ $2y = -x$ ⟺ $y = -\frac{1}{2}x + 0$; The slope is $-\frac{1}{2}$ and
the y-intercept is 0. See Figure 25.

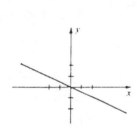 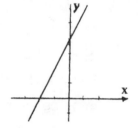

Figure 25   Figure 28

28) $x + 2 = \frac{1}{2}y$ ⟺ $y = 2x + 4$; The slope is 2 and the y-intercept is
4. See Figure 28.

31) Think of this problem as having a <u>time</u> variable and a <u>price</u> variable
where the price is dependent on the time. When $t = 0$, $p = 59{,}000$.
When $t = 6$, $p = 95{,}000$. This is equivalent to having the points
$(0, 59{,}000)$ and $(6, 95{,}000)$ on a graph. The slope is

$\dfrac{95,000 - 59,000}{6 - 0} = 6,000$. The interpretation of the slope is that

the price has risen $6,000 per year if the linear assumption is valid. Using the Slope-Intercept Form with the point (0, 59,000), we have p = 6000t + 59,000, where p is the value of the house and t is the number of years after the purchase date; { This could also be written as y = 6000x + 59,000. } The house was worth $73,000 when p was equal to 73,000. 73,000 = 6000t + 59,000 $\Rightarrow$ 6000t = 14,000 $\Rightarrow$ t = $\frac{7}{3}$ years or 2 years and 4 months after the purchase date.

34) Use F = mC + b with points (0, 32) and (100, 212). Substituting (0, 32) we obtain 32 = b. Now substitute (100, 212) into F = mC + 32 to obtain 212 = 100m + 32 $\Rightarrow$ 180 = 100m $\Rightarrow$ m = $\frac{9}{5}$; F = $\frac{9}{5}$C + 32; If C increases by 1^0, F will increase by $(\frac{9}{5})^0$.

37) (a) The monthly take-home pay (M) is equal to the monthly revenue (R) minus $500 to the parent company minus $400 in fixed costs minus the 50% of the revenue that is not profit minus the 10% of the profits which also goes to the parent company. Writing this in equation form, we have

M = R - 500 - 400 - 50% (R) - 10% (50% (R))

$\quad$ = R - 900 - .5R - .05R = (1 - 0.55) R - 900

$\quad$ = 0.45 R - 900.

(b) To break even would mean a take-home pay of $0. Setting M = 0 in part (a) yields 0 = 0.45 R - 900 $\Longleftrightarrow$ 900 = $\frac{45}{100}$R $\Longleftrightarrow$ R = 20 (100) = $2,000 per month.

Exercises 1.5

40) (a) The slope of the line from $(-4, -3)$ to the origin is $\frac{3}{4}$, so the slope of the tangent line (which is perpendicular to the line to the origin) is $-\frac{4}{3}$. The line through $(-4, -3)$ with a slope of $-\frac{4}{3}$ is $4x + 3y = -25$. Letting $y = -50$ and solving for x we obtain $x = 31.25$.

(b) The slope of the line from (x, y) to the origin is $\frac{y}{x}$, so the slope of the tangent line is $-\frac{x}{y}$. The line through $(0, -50)$ is $y + 50 = \left(-\frac{x}{y}\right)(x - 0)$ or $y^2 + 50y = -x^2$. The equation of the circle is $x^2 + y^2 = 25$ or $-x^2 = y^2 - 25$. Substituting that expression into the equation of the line yields $y^2 + 50y = y^2 - 25$ $\Rightarrow$ $50y = -25$ $\Rightarrow$ $y = -\frac{1}{2}$. If $y = -\frac{1}{2}$, then $x^2 + \frac{1}{4} = 25$ $\Rightarrow$ $x^2 = \frac{99}{4}$ $\Rightarrow$ $x = \pm\frac{3\sqrt{11}}{2}$ $\{\approx \pm 4.975\}$. We want the negative x value according to the diagram.

Exercises 1.6 (Review)

1) (a) $-0.1 < -0.01$ (b) $\sqrt{9} > -3$ {since $\sqrt{9} = 3$}   (c) $\frac{1}{6} > 0.166$
   (c) $\frac{1}{6} > 0.166$ {since $\frac{1}{6} = 0.16\overline{6}$}

4) (a) $d(A, C) = |(-3) - (-8)| = |5| = 5$ (b) $d(C, A) = |(-8) - (-3)|$
   $= |-5| = -(-5) = 5$ (c) $d(B, C) = |(-3) - (4)| = |-7| = 7$

7) The surface area formula for a cylinder is $S = 2\Pi rh + 2\Pi r^2$. Letting $S = 10\Pi$ and $h = 4$, we obtain $10\Pi = 2\Pi r^2 + 8\Pi r$ $\Rightarrow$ $5 = r^2 + 4r$ $\Rightarrow$ $(r + 5)(r - 1) = 0$ $\Rightarrow$ $r = 1$. The diameter is twice the radius or 2 feet.

10) $\dfrac{6r^3y^2}{2r^5y} = \dfrac{3y}{r^2}$

13) $\sqrt[3]{(x^4y^{-1})^6} = (x^4y^{-1})^{6/3} = (x^4y^{-1})^2 = x^8y^{-2} = \dfrac{x^8}{y^2}$

16) $\sqrt{\dfrac{a^2b^3}{c}} = \sqrt{\dfrac{a^2b^3}{c} \cdot \dfrac{c}{c}} = \sqrt{\dfrac{a^2b^2}{c^2}} \sqrt{\dfrac{bc}{1}} = \dfrac{ab\sqrt{bc}}{c}$

19) $2c^3 - 12c^2 + 3c - 18 = 2c^2(c - 6) + 3(c - 6) = (2c^2 + 3)(c - 6)$

22) $x^4 - 8x^3 + 16x^2 = x^2(x^2 - 8x + 16) = x^2(x - 4)^2$

25) $(3y^3 - 2y^2 + y + 4)(y^2 - 3) = (3y^3 - 2y^2 + y + 4)(y^2) +$

$(3y^3 - 2y^2 + y + 4)(-3) = 3y^5 - 2y^4 + y^3 + 4y^2 - 9y^3 + 6y^2 - 3y -$

$12 = 3y^5 - 2y^4 - 8y^3 + 10y^2 - 3y - 12$

28) $\dfrac{r^3 - t^3}{r^2 - t^2} = \dfrac{(r - t)(r^2 + rt + t^2)}{(r - t)(r + t)} = \dfrac{r^2 + rt + t^2}{r + t}$

31) $\dfrac{7}{x + 2} + \dfrac{3x}{(x + 2)^2} - \dfrac{5}{x} = \dfrac{7(x)(x + 2) + 3x(x) - 5(x + 2)^2}{x(x + 2)^2} =$

$\dfrac{7x^2 + 14x + 3x^2 - 5x^2 - 20x - 20}{x(x + 2)^2} = \dfrac{5x^2 - 6x - 20}{x(x + 2)^2}$

34) $10 - 7x < 4 + 2x \Rightarrow -9x < -6 \Rightarrow x > \tfrac{2}{3}; \ (\tfrac{2}{3}, \infty)$

37) $10x^2 + 11x > 6 \Rightarrow 10x^2 + 11x - 6 > 0 \Rightarrow (2x + 3)(5x - 2) > 0$

Sign of the product $\qquad\qquad\ + \ | \ - \ | \ +$
Sign of the factor $(5x - 2) \ - - - | - - - | + + +$
Sign of the factor $(2x + 3) \ \underline{\ - - - | + + + | + + +}$
$\qquad\qquad\qquad\qquad\qquad\quad -3/2 \qquad 2/5$

The product $(2x + 3)(5x - 2)$ is greater than 0 on the intervals

$(-\infty, -\tfrac{3}{2})$ and $(\tfrac{2}{5}, \infty); \ (-\infty, -\tfrac{3}{2}) \cup (\tfrac{2}{5}, \infty)$

40) $L < \tfrac{1}{2}L_0 \Rightarrow L^2 < \tfrac{1}{4}L_0^2 \ \{\text{squaring both sides}\} \Rightarrow$

$L_0^2\left[1 - \dfrac{V^2}{c^2}\right] < \tfrac{1}{4}L_0^2 \ \{\text{substituting the given equation}\} \Rightarrow$

$\left[1 - \dfrac{V^2}{c^2}\right] < \tfrac{1}{4} \Rightarrow \dfrac{c^2 - V^2}{c^2} < \tfrac{1}{4} \Rightarrow 4c^2 - 4V^2 < c^2$

$\Rightarrow 4V^2 > 3c^2 \Rightarrow V > \dfrac{\sqrt{3}\,c}{2} \ \{c \text{ is positive}\}$

43) If $\frac{y}{x} < 0$, then y and x must be opposite in sign. The points that have coordinates that are opposite in sign are in quadrants II and IV, so these two quadrants make up the solution set.

46) The general equation of a circle with center C $(7, -4)$ is $(x - 7)^2 + (x + 4)^2 = r^2$. $x = -3$ and $y = 3$ must satisfy that equation; $(-10)^2 + (7)^2 = r^2 \Rightarrow r^2 = 149$, so the equation of the circle is $(x - 7)^2 + (x + 4)^2 = 149$.

49) (a) $6x + 2y + 5 = 0 \iff 2y = -6x - 5 \iff y = -3x - \frac{5}{2}$; The slope of the given line is -3, so the slope of the required line is also -3. $y + \frac{1}{3} = -3 (x - \frac{1}{2}) \Rightarrow 6y + 2 = -18x + 9 \Rightarrow 18x + 6y - 7 = 0$

(b) The required slope is now $\frac{1}{3}$. $y + \frac{1}{3} = \frac{1}{3} (x - \frac{1}{2}) \Rightarrow 6y + 2 = 2x - 1 \Rightarrow 2x - 6y - 3 = 0$

52) The graph is a line with x-intercept 4 and y-intercept $-\frac{3}{4}$. See Figure 52.

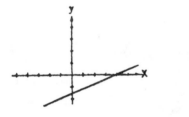

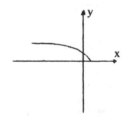

Figure 52   Figure 55

55) The graph is a half-parabola with x-intercept 1 and y-intercept 1. See Figure 55.

58) The graph is a circle with
center at the origin and
radius 4. $(x^2 + y^2 = 16)$
See Figure 58.

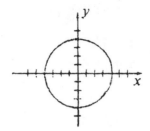

Figure 58

Exercises 2.1

1) Given $f(x) = 2x^2 - 3x + 4$, we have the following :

$f(1) = 2(1)^2 - 3(1) + 4 = 2 - 3 + 4 = 3$

$f(-1) = 2(-1)^2 - 3(-1) + 4 = 2 + 3 + 4 = 9$

$f(0) = 2(0)^2 - 3(0) + 4 = 0 + 0 + 4 = 4$

$f(2) = 2(2)^2 - 3(2) + 4 = 8 - 6 + 4 = 6$

4) Given $f(x) = \dfrac{x}{x - 2}$, we have the following : $f(1) = \dfrac{1}{1 - 2} = -1$;

$f(3) = \dfrac{3}{3 - 2} = 3$; $f(-2) = \dfrac{-2}{-2 - 2} = \dfrac{1}{2}$; $f(0) = \dfrac{0}{0 - 2} = 0$

7) Given $f(x) = 2x^2 - x + 3$, we have the following :

(a) $f(a) = 2a^2 - a + 3$

(b) $f(-a) = 2(-a)^2 - (-a) + 3 = 2a^2 + a + 3$

(c) $-f(a) = -1 \cdot f(a) = -1(2a^2 - a + 3) = -2a^2 + a - 3$

(d) $f(a + h) = 2(a + h)^2 - (a + h) + 3 = 2a^2 + 4ah + 2h^2 - a - h + 3$

(e) $f(a) + f(h) = (2a^2 - a + 3) + (2h^2 - h + 3) =$

$2a^2 - a + 2h^2 - h + 6$

(f) Use part (d) for $f(a + h)$ and part (a) for $f(a)$.

$\dfrac{f(a + h) - f(a)}{h} = \dfrac{(2a^2 + 4ah + 2h^2 - a - h + 3) - (2a^2 - a + 3)}{h} =$

$\dfrac{4ah + 2h^2 - h}{h} = \dfrac{h(4a + 2h - 1)}{h} = 4a + 2h - 1$

10) Given $g(x) = 3x - 8$, we have the following :

(a) $g\left(\dfrac{1}{a}\right) = 3\left(\dfrac{1}{a}\right) - 8 = \dfrac{3}{a} - 8 = \dfrac{3 - 8a}{a}$

(b) $\dfrac{1}{g(a)} = \dfrac{1}{3(a) - 8} = \dfrac{1}{3a - 8}$

(c) $g(a^2) = 3(a^2) - 8 = 3a^2 - 8$

(d) $(g(a))^2 = (3a - 8)^2$ or $9a^2 - 48a + 64$

(e) $g(\sqrt{a}) = 3(\sqrt{a}) - 8 = 3\sqrt{a} - 8$

(f) $\sqrt{g\,(a)} = \sqrt{3a - 8}$

13) The domain of $f\,(x) = \sqrt{3x - 5}$ is the set of values that make the radicand ≥ 0; i.e. $3x - 5 \geq 0 \Rightarrow x \geq \frac{5}{3}$; $[\frac{5}{3}, \infty)$

16) The domain of $f\,(x) = \sqrt{x^2 - 9}$ is the set of values that make the radicand ≥ 0; i.e. $x^2 - 9 \geq 0 \Rightarrow x^2 \geq 9 \Rightarrow |x| \geq 3 \Rightarrow$ $x \geq 3$ or $x \leq -3$; $(-\infty, -3] \cup [3, \infty)$

19) For this function we must have the radicand ≥ 0 <u>and</u> the denominator cannot equal 0. The radicand is ≥ 0 if $x \geq 0$. The denominator is $(2x - 3)\,(x - 4)$, so $x \neq \frac{3}{2}$ or 4. The solution could be written as all nonnegative real numbers except $\frac{3}{2}$ and 4. In interval notation, we have $[0, \frac{3}{2}) \cup (\frac{3}{2}, 4) \cup (4, \infty)$.

22) The domain is $x \geq 0$ as in Exercise 19. Since the result of a square root will be at least 0, the range will consist of numbers that are at least 2. Domain : $[0, \infty)$ Range : $[2, \infty)$ See Figure 22.

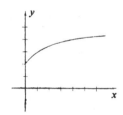

Figure 22   Figure 25

25) The domain will be all values except 0 since 0 makes the function undefined. The range will be all values except 0 since the fraction can take on any nonzero value. i.e. Let c be some value. $c = \frac{1}{x} \Rightarrow$ $x = \frac{1}{c}$ which can be found for any c not equal to zero. Domain : $(-\infty, 0) \cup (0, \infty)$ Range : $(-\infty, 0) \cup (0, \infty)$ See Figure 25.

Exercises 2.1

28) $f(-x) = 7(-x)^6 - (-x)^4 + 7 = 7x^6 - x^4 + 7 = f(x)$ so f is even.

31) $f(-x) = 2 = f(x)$ so f is even.

34) $f(-x) = \sqrt{(-x)^2 + 1} = \sqrt{x^2 + 1} = f(x)$ so f is even.

37) The formula for the volume of a box is $V = \ell w h$. { Volume = length X width X height } In this case, the length is $30 - 2x$ { since x is taken off each end }, the width is $20 - 2x$, and the height is x.

$V = \ell w h = (30 - 2x)(20 - 2x)(x) = 4x^3 - 100x^2 + 600x$

40) (a) The formulas used here are $V = \frac{4}{3}\Pi r^3$ { for the volume of a sphere } and $S = 4\Pi r^2$ { for the surface area of a sphere }. Solving the volume formula for r yields the following :

$$V = \frac{4}{3}\Pi r^3 \;\Rightarrow\; r^3 = \frac{3V}{4\Pi} = \sqrt[3]{\frac{3V}{4\Pi}}\;;\;\text{Substituting that value into the}$$

surface area formula yields : $S = 4\Pi\left(\sqrt[3]{\frac{3V}{4\Pi}}\right)^2 =$

$$\frac{(4\Pi)(3V)^{2/3}}{(4\Pi)^{2/3}} = (4\Pi)^{1/3}(3V)^{2/3} = \left[(4\Pi)^1(3V)^2\right]^{1/3} = \sqrt[3]{36V^2\Pi}$$

(b) $S = \sqrt[3]{36(12)^2\,\Pi} = \sqrt[3]{3(12)^3\,\Pi} = 12\sqrt[3]{3\,\Pi}$ { ≈ 25.348 }

43) (a) CTP forms a right angle so the Pythagorean Formula may be applied. $r^2 + y^2 = (h + r)^2 \;\Rightarrow\; y^2 = (h^2 + 2hr + r^2) - r^2 \;\Rightarrow$
$y^2 = h^2 + 2hr \;\Rightarrow\; y = \sqrt{h^2 + 2hr}$

(b) $y = \sqrt{(200)^2 + 2(200)(4000)} = \sqrt{1640000} = 200\sqrt{41}$
$\approx 1{,}280.6$ miles

46) This is a function which can have two values. The first : If less than 50 pairs are ordered, A will simply be ($20) (x). The second : If 50 to 600 pairs are ordered, each pair is reduced by 2 cents. This is represented by 0.02x (the discount amount). The cost of a pair of shoes is then represented by $20 - 0.02x$. The amount A is

the cost of a pair of shoes times the number of pairs, which is (20 − 0.02x) (x). This can be summarized with the following form :

$$A(x) = \begin{cases} 20x & \text{if } x < 50 \\ (20 - 0.02x)(x) & \text{if } 50 \leq x \leq 600 \end{cases}$$

49) (a) Form one triangle with the base 4 and height 12. Form another triangle with the base 4 − r {the distance from the cylinder to the cone} and height h. Using similar triangle proportions, we have $\dfrac{h}{4-r} = \dfrac{12}{4} \Rightarrow 4h = 12(4-r) \Rightarrow h = 3(4-r)$.

(b) $V = \Pi r^2 h = \Pi r^2 (3(4-r)) = 3\Pi r^2 (4-r)$

52) This is a non-vertical line. For each x value there is exactly one ordered pair (x, y).

55) This is the same solution as above in Exercise 52.

58) This is the same solution as above in Exercise 52.

60) {For an example that does not satisfy the definition.} The set of points in this problem is the half-plane under the line y = x. Take the points (7, −3) and (7, 4). In both cases, y < x. So for each x value there is <u>not</u> exactly one ordered pair (x, y) having x in the first position.

1) Domain : (−∞, ∞);
   Range : (−∞, ∞);
   increasing on (−∞, ∞);
   See Figure 1.

Figure 1

<u>Exercises 2.2</u>

4) Domain : (-∞, ∞); Range : (-∞, ∞); increasing on (-∞, ∞); See
   Figure 4.

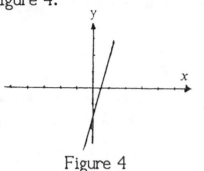

Figure 4   Figure 7

7) Domain : (-∞, ∞); Since $2x^2$ is always ≥ 0, the y values will always
   be ≥ -4; Range : [-4, ∞); decreasing on (-∞, 0]; increasing on
   [0, ∞); See Figure 7.

10) 4 - x must be ≥ 0 which implies that x ≤ 4; Domain : (-∞, 4];
    The result of the square root is always nonnegative. Range : [0, ∞);
    decreasing on (-∞, 4]; See Figure 10.

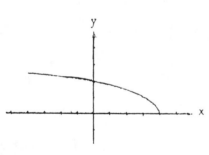
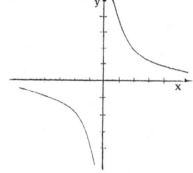

Figure 10   Figure 13

13) x = 0 would make the function undefined; Domain :
    (-∞, 0) U (0, ∞); y can be any nonzero number; Range :
    (-∞, 0) U (0, ∞); decreasing on (-∞, 0) and on (0, ∞); See Fig.13

16) Domain : (-∞, ∞); y can only be nonnegative; Range : [0, ∞);
    decreasing on (-∞, -2]; increasing on [-2, ∞); See Figure 16.

24

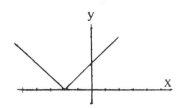

Figure 16

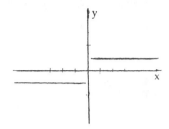

Figure 19

19) Domain : $(-\infty, 0) \cup (0, \infty)$; Since $|x|$ is equal to x or $-x$, the quotient will either be 1 or -1. Range : $\{-1, 1\}$; constant on $(-\infty, 0)$ and on $(0, \infty)$; See Figure 19.

22) The graphs are lines with a slope of -2. The y-intercepts are 0, 1, and -3. See Figure 22.

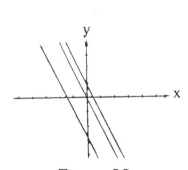

Figure 22

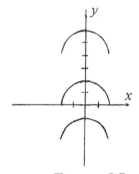

Figure 25

25) The graphs are semicirles. The c value shifts the graph of $f(x) = \sqrt{4 - x^2}$ vertically c units. In this case, 0 units, then 4 units up, and finally 3 units down. See Figure 25.

28) The graphs are parabolas opening down. The c value shifts the graph of $f(x) = -2x^2$ horizontally c units. In this case, 0 units, then 3 units right and finally 1 unit right. See Figure 28.

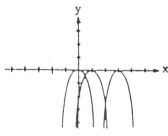

Figure 28

31) (a) shift left 2 units

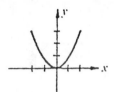

(b) shift right 2 units

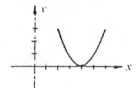

(c) shift up 2 units

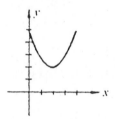

(d) shift down 2 units

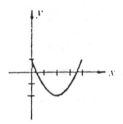

(e) multiply the y values by 2

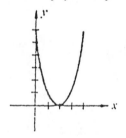

(f) multiply the y values by $\frac{1}{2}$

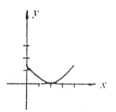

(g) multiply the y values by -2

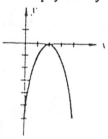

(h) shift right 3, then shift up 1

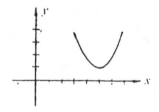

34) When x is an integer, y will be -1. This has the effect of "making little holes" in the line y = 1. See Figure 34.

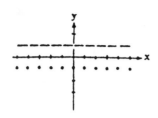

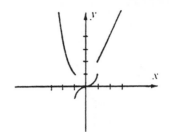

Figure 34   Figure 37

37) We want the parabola $y = x^2$ when $x \le -1$, the cubic $y = x^3$ when x is between -1 and 1 $\{$ since $|x| < 1 \Rightarrow -1 < x < 1 \}$, and the line $y = 2x$ when $x \ge 1$. See Figure 37.

40) $\dfrac{x^2 - 1}{1 - x} = \dfrac{(x + 1)(x - 1)}{-(x - 1)} = -(x + 1)$

provided that $x \ne 1$ since that would make the original function undefined; This is just the line $y = -x - 1$ with a slope of -1 and y-intercept of -1. When x <u>is</u> equal to 1, the function is defined to be 2. See Figure 40.

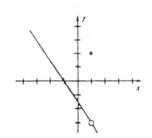

Figure 40

1) The graphs are all parabolas with y-intercept $(0, 2)$.

   (a)   (b)

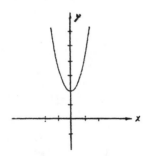

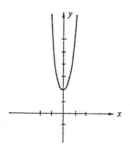

(c)

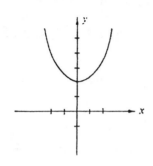

(d)

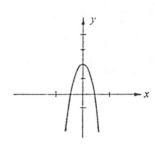

4) Since there is no x term in this function, the vertex is on the y-axis. Its y-coordinate is equal to the constant term.

Vertex : (0, 3); See Figure 4.

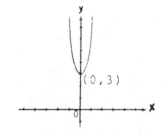

(0, 3)

Figure 4

7) $\{a = 4, b = -11, c = -3\}$ Substituting these values into the

Quadratic Formula yields : $x = \dfrac{11 \pm \sqrt{121 + 48}}{8} = \dfrac{11 \pm 13}{8} = 3, -\dfrac{1}{4}$

10) $\{a = 10, b = 1, c = -21\}$ Substituting these values into the

Quadratic Formula yields : $x = \dfrac{-1 \pm \sqrt{1 + 840}}{20} = \dfrac{-1 \pm 29}{20} =$

$\dfrac{7}{5}, -\dfrac{3}{2}$

13) $f(x) = -5x^2 - 10x + 3 \Rightarrow f(x) = -5(x^2 + 2x) + 3 \Rightarrow$

$f(x) = -5(x^2 + 2x + 1) + 3 + 5$ { add 5 since we really subtracted 5 when we added 1 inside the parentheses } $\Rightarrow$

$f(x) = -5(x + 1)^2 + 8$; Since $a = -5$ is negative, the parabola opens down and we have a maximum of 8 when $x = -1$.

16) $f(x) = x^2 - 6x = x(6 - x)$; There are x-intercepts at 0 and 6.

Completing the square to find the vertex yields :

$f(x) = x^2 - 6x + 9 - 9 \Rightarrow f(x) = (x - 3)^2 - 9 \Rightarrow$ the vertex is at $(3, -9)$; See Figure 16.

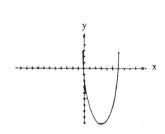

Figure 16

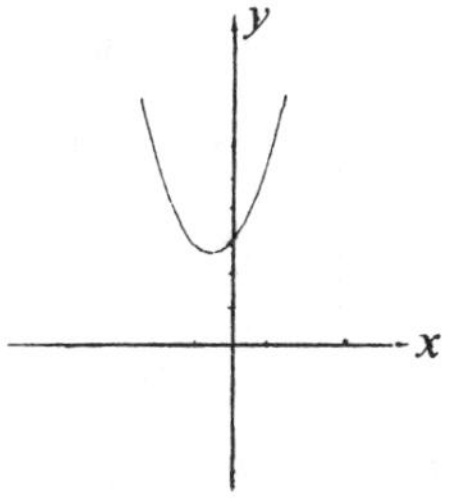

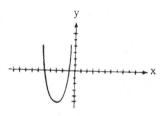

Figure 19

19) To find the vertex : $f(x) = (x^2 + x + \frac{1}{4}) + 3 - \frac{1}{4} \Rightarrow$

$f(x) = (x + \frac{1}{2})^2 + \frac{11}{4} \Rightarrow V(-\frac{1}{2}, \frac{11}{4})$; Since the parabola opens up and the vertex is above the x-axis, there are no x-intercepts. The y-intercept is 3. See Figure 19.

22) The y-intercept is 13. To find the

vertex : $f(x) = 2(x^2 + 6x) + 13$

$\Rightarrow f(x) = 2(x^2 + 6x + 9) + 13 - 18$

$\Rightarrow f(x) = 2(x + 3)^2 - 5 \Rightarrow$

$V(-3, -5)$; The x-intercepts :

$x = \dfrac{-12 \pm \sqrt{144 - 104}}{4}$

$= -3 \pm \frac{1}{2}\sqrt{10}$ See Figure 22.

Figure 22

25) Treat this as a parabola with vertex $(\frac{9}{2}, 3)$. The general equation of a parabola with the vertex at $(\frac{9}{2}, 3)$ is $y = a(x - \frac{9}{2})^2 + 3$. Since $(0, 0)$ is on the parabola, we can substitute these values for x and y in the general equation. $0 = a(-\frac{9}{2})^2 + 3 \Rightarrow -3 = a(\frac{81}{4}) \Rightarrow$

Exercises 2.3

$a = -\frac{4}{27}$. The equation is now $y = -\frac{4}{27}(x - \frac{9}{2})^2 + 3$. This can also be written as $y = \frac{4}{27}x(9 - x)$. {with some algebraic manipulation}

28) (a) Substituting $x = 0$ and $m = 0$ into $m = 2ax + b$ yields $b = 0$. Substituting $x = 800$ and $m = \frac{1}{5}$ into $m = 2ax$ yields $a = \frac{1}{8000}$. The equation is $y = \frac{1}{8000}x^2$. {$c = 0$ since the parabola passes through the origin}

   (b) $y = \frac{1}{8000}(800)^2 = 80$; The coordinates are $(800, 80)$.

31) The height of the building can be found by letting $t = 0$. This yields $s(t) = 100$ feet. To find the maximum height, we complete the square. $s(t) = -16t^2 + 144t + 100 = -16(t^2 - 9t) + 100 = -16(t^2 - 9t + \frac{81}{4}) + 100 + 324 = -16(t - \frac{9}{2})^2 + 424$; This result indicates that the maximum height of 424 feet occurs after $4\frac{1}{2}$ sec.

34) Completing the square yields this form of the given equation : $y = -\frac{1}{30}(v - \frac{75}{2})^2 + \frac{375}{8}$; This indicates that a speed of 37.5 mph results in a mileage rating of 46.875 miles per gallon.

37) (a) The formula for the perimeter of a rectangle is $P = 2x + 2y$. Now $P = 24$, so $24 = 2x + 2y \Rightarrow 12 = x + y \Rightarrow y = 12 - x$.

   (b) Area = length X width = $yx = (12 - x)x$

   (c) Complete the square for $A = -x^2 + 12x$ (from part (b)). $A = -1(x^2 - 12x) = -1(x^2 - 12x + 36) + 36 = -(x - 6)^2 + 36$; This indicates that the area is at a maximum when $x = 6$. Since $y = 12 - x$, y is also 6 and the rectangle is a square.

40) (a) Let y denote the number of $1 decreases in the monthly charge.

$\quad$ R (y) = (# of customers) (monthly charge)

$\qquad$ = (5000 + 500y) (20 − y) (⋆)

$\qquad$ = −500y² + 5000y + 100000

$\qquad$ = −500 (y − 5)² + 112500

Now let x denote the monthly charge which is 20 − y.

(⋆) becomes (5000 + 500 (20 − x)) (x) or 500x (30 − x).

(b) From part (a), y = 5 results

in a $15 monthly charge and

7500 customers for a revenue

of $112,500. See Figure 40.

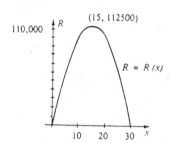

Figure 40

1) (f + g) (x) = f (x) + g (x) = 3x² + $\dfrac{1}{2x - 3}$;

$\quad$ (f − g) (x) = f (x) − g (x) = 3x² − $\dfrac{1}{2x - 3}$;

$\quad$ (f g) (x) = f (x) · g (x) = (3x²) · $\left(\dfrac{1}{2x - 3}\right)$ = $\dfrac{3x^2}{2x - 3}$;

$\quad$ $\left(\dfrac{f}{g}\right)$ (x) = $\dfrac{f (x)}{g (x)}$ = $\dfrac{3x^2}{\dfrac{1}{2x - 3}}$ = 3x² (2x − 3)

4) (f + g) (x) = f (x) + g (x) = (x³ + 3x) + (3x² + 1) = x³ + 3x² + 3x + 1;

$\quad$ (f − g) (x) = f (x) − g (x) = (x³ + 3x) − (3x² + 1) = x³ − 3x² + 3x − 1;

$\quad$ (f g) (x) = f (x) · g (x) = (x³ + 3x) (3x² + 1) = 3x⁵ + 10x³ + 3x;

Exercises 2.4

$$\left(\frac{f}{g}\right)(x) = \frac{f(x)}{g(x)} = \frac{x^3 + 3x}{3x^2 + 1}$$

7) $(f \circ g)(x) = f(g(x)) = f(2x - 1) = 3(2x - 1) + 2 = 6x - 1;$

$\quad (g \circ f)(x) = g(f(x)) = g(3x + 2) = 2(3x + 2) - 1 = 6x + 3$

10) $(f \circ \dot{g})(x) = f(g(x)) = f(2x^2) = 7(2x^2) + 1 = 14x^2 + 1;$

$\quad (g \circ f)(x) = g(f(x)) = g(7x + 1) = 2(7x + 1)^2 = 98x^2 + 28x + 2$

13) $(f \circ g)(x) = f(g(x)) = f(x^3) = x^3 - 1;$

$\quad (g \circ f)(x) = g(f(x)) = g(x - 1) = (x - 1)^3 = x^3 - 3x^2 + 3x - 1$

16) $(f \circ g)(x) = f(g(x)) = f(x^3 + 1) = \sqrt[3]{(x^3 + 1)^2 + 1} = \sqrt[3]{x^6 + 2x^3 + 2};$

$\quad (g \circ f)(x) = g(f(x)) = g(\sqrt[3]{x^2 + 1}) = (\sqrt[3]{x^2 + 1})^3 + 1 = x^2 + 2$

19) $(f \circ g)(x) = f(g(x)) = f(-5) = |-5| = 5;$

$\quad (g \circ f)(x) = g(f(x)) = g(|x|) = -5 \quad \{g(\text{anything}) = -5\}$

22) $(f \circ g)(x) = f(g(x)) = f(x + 1) = \dfrac{1}{(x + 1) + 1} = \dfrac{1}{x + 2};$

$\quad (g \circ f)(x) = g(f(x)) = g\left(\dfrac{1}{x + 1}\right) = \dfrac{1}{x + 1} + 1 = \dfrac{1 + (x + 1)}{x + 1} = \dfrac{x + 2}{x + 1}$

25) $r = f(t) = 6t$ and $A = g(r) = \Pi r^2;$ We want the composite function

$\quad (g \circ f)(t).$ $(g \circ f)(t) = g(f(t)) = g(6t) = \Pi (6t)^2 = 36 \Pi t^2$ sq ft/min

28) For one face of the cube we have $x^2 + x^2 = y^2$ by the Pythagorean

Formula. $y^2 = 2x^2 \Rightarrow y = \sqrt{2}\,x;$ Now consider the right triangle

formed by a diagonal of a face and one of the edges. $y^2 + x^2 = d^2 \Rightarrow$

$d = \sqrt{y^2 + x^2} = \sqrt{2x^2 + x^2} = \sqrt{3x^2} = \sqrt{3}\,x$

Exercises 2.5

1) If $a \neq b$, then $2a \neq 2b$ and hence $2a + 9 \neq 2b + 9$, or $f(a) \neq f(b)$.
Thus, f is one-to-one.

4) An equivalent condition to showing if $a \neq b$, then $f(a) \neq f(b)$ is to
show if $f(a) = f(b)$, then $a = b$. We will show this condition here.

First, complete the square for f. $f(x) = 2x^2 - x - 3$

$= 2(x^2 - \frac{1}{2}x) - 3 = 2(x^2 - \frac{1}{2}x + \frac{1}{16}) - 3 - \frac{1}{8} = 2(x - \frac{1}{4})^2 - \frac{25}{8}$; Now

suppose $f(a) = f(b)$. Then $2(a - \frac{1}{4})^2 - \frac{25}{8} = 2(b - \frac{1}{4})^2 - \frac{25}{8}$ $\longleftrightarrow$

$2(a - \frac{1}{4})^2 = 2(b - \frac{1}{4})^2$ $\longleftrightarrow$ $(a - \frac{1}{4})^2 = (b - \frac{1}{4})^2$ $\longleftrightarrow$

$(a - \frac{1}{4}) = \pm(b - \frac{1}{4})$ $\longleftrightarrow$ $a - \frac{1}{4} = b - \frac{1}{4}$ $\underline{or}$ $a - \frac{1}{4} = -b + \frac{1}{4}$; The first

equation implies that $a = b$ but the second equation does not yield that

result. Thus, f is $\underline{not}$ one-to-one. In particular, $f(0) = f(\frac{1}{2}) = -3$.

7) As in Exercise 4, suppose $f(a) = f(b)$. Then $|a| = |b|$. Using the

definition of absolute value, we have $\pm a = \pm b$, and cannot conclude

that $a = b$. Thus, f is $\underline{not}$ one-to-one. In particular, $f(-1) = f(1) = 1$.

10) If $a \neq b$, then $\sqrt[3]{a} \neq \sqrt[3]{b}$ or $f(a) \neq f(b)$. Thus, f is one-to-one.

For Exercises 13 & 16 we must verify that $(f \circ g)(x) = (g \circ f)(x) = x$.

13) $(f \circ g)(x) = f(g(x)) = f\left[\frac{x - 5}{7}\right] = 7\left[\frac{x - 5}{7}\right] + 5 = (x - 5) + 5 = x$;

$(g \circ f)(x) = g(f(x)) = g(7x + 5) = \frac{(7x + 5) - 5}{7} = \frac{7x}{7} = x$; Thus, f

and g are inverse functions. See Figure 13.

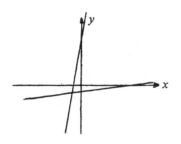

Figure 13

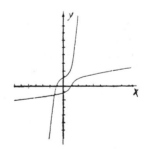

Figure 16

16) $(f \circ g)(x) = f(g(x)) = f(\sqrt[3]{x - 1}) = (\sqrt[3]{x - 1})^3 + 1 = (x - 1) + 1 = x$;

$(g \circ f)(x) = g(f(x)) = g(x^3 + 1) = \sqrt[3]{(x^3 + 1) - 1} = \sqrt[3]{x^3} = x$; Thus, f

and g are inverse functions. See Figure 16.

Exercises 2.5

19) Given $y = \dfrac{1}{2x + 5}$, exchange x with y. Now $x = \dfrac{1}{2y + 5}$ and we want

to solve this equation for y. $x(2y + 5) = 1 \Rightarrow 2xy + 5x = 1 \Rightarrow$

$(2x)y = 1 - 5x \Rightarrow y = \dfrac{1 - 5x}{2x} = f^{-1}(x)$. For f the domain is

$x > -\frac{5}{2}$. This would make the original denominator positive and

thus the entire fraction positive so that the range of f is $y > 0$. For

f^{-1}, the domain equals the range of f. i.e. Domain of f^{-1} is $x > 0$.

22) Given $y = 4x^2 + 1$, exchange x with y. Now $x = 4y^2 + 1$ and we

want to solve this equation for y. $x - 1 = 4y^2 \Rightarrow y^2 = \dfrac{x - 1}{4} \Rightarrow$

$y = \pm\sqrt{\dfrac{x - 1}{4}}$; Since $x \geq 0$ is the domain of f, $y \geq 0$ is the range of

f^{-1}. Thus we select the "+" instead of the "-". If $x \geq 0$ for f, $y \geq 1$.

The domain of f^{-1} is the range of f, so the domain of f^{-1} is $x \geq 1$

where $f^{-1}(x) = \sqrt{\dfrac{x - 1}{4}}$. Note that if the domain of f was $x \leq 0$, we

would have selected the "-" for f^{-1}.

25) Given $y = \sqrt{3x - 5}$, exchange x with y. Now $x = \sqrt{3y - 5}$ and we

want to solve this equation for y. $x^2 = 3y - 5 \Rightarrow x^2 + 5 = 3y \Rightarrow$

$y = \dfrac{x^2 + 5}{3}$. For f the domain is $x \geq \frac{5}{3}$ and the range is $y \geq 0$. Thus

for $f^{-1}(x) = \dfrac{x^2 + 5}{3}$, the domain is $x \geq 0$ and the range is $y \geq \frac{5}{3}$.

28) Given $y = (x^3 + 1)^5$, exchange x with y. Now $x = (y^3 + 1)^5$ and we

want to solve this equation for y. $\sqrt[5]{x} = (y^3 + 1) \Rightarrow \sqrt[5]{x} - 1 = y^3$

$\Rightarrow y = f^{-1}(x) = \sqrt[3]{\sqrt[5]{x} - 1}$; The domain and range for both f and f^{-1}

is the set of all real numbers.

31) (a) Since f is one-to-one, an inverse exists. To find the inverse, exchange x with y in $y = ax + b$. $x = ay + b \Rightarrow x - b = ay \Rightarrow$

$y = f^{-1}(x) = \dfrac{x - b}{a}$ since $a \neq 0$.

(b) No. If f (x) is a constant function, then the equation in part (a) would have a = 0, a contradiction. Geometrically, the inverse of a horizontal line is a vertical line which is not a function. Also, a constant function is not one-to-one.

34) (a) If P (a, b) is on the graph of f, then f (a) = b. Also $f^{-1}(f(a)) = f^{-1}(b)$ or $a = f^{-1}(b)$. So, Q (b, a) is on the graph of f^{-1}.

(b) The midpoint of PQ is $\left(\dfrac{a + b}{2}, \dfrac{b + a}{2} \right)$. Since the x and y coordinates are equal, the point is on the line y = x.

(c) The slope between P and Q is $m_{PQ} = \dfrac{a - b}{b - a} = \dfrac{-1\,(b - a)}{b - a} = -1$

and the slope of the line y = x is 1. Since the slopes of the lines are negative reciprocals of each other, the lines are perpendicular.

37) (a) Domain of f : [−3, 3];
Range of f : [−2, 2];
Domain of f^{-1} : [−2, 2];
Range of f^{-1} : [−3, 3];
See Figure 37.

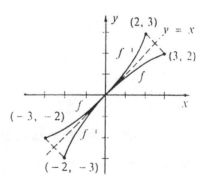

Figure 37

1) (a) $f(1) = \dfrac{1}{\sqrt{1+3}} = \dfrac{1}{2}$
(b) $f(-1) = \dfrac{-1}{\sqrt{-1+3}} = \dfrac{-1}{\sqrt{2}} = -\dfrac{\sqrt{2}}{2}$

(c) $f(0) = \dfrac{0}{\sqrt{0+3}} = 0$
(d) $f(-x) = \dfrac{-x}{\sqrt{-x+3}} = \dfrac{-x}{\sqrt{3-x}}$

(e) $-f(x) = -1\left(\dfrac{x}{\sqrt{x+3}}\right) = \dfrac{-x}{\sqrt{x+3}}$
(f) $f(x^2) = \dfrac{x^2}{\sqrt{x^2+3}}$

(g) $(f(x))^2 = \left(\dfrac{x}{\sqrt{x+3}}\right)^2 = \dfrac{x^2}{x+3}$

4) The domain of $f(x) = |x+3|$ is all reals since any value can be substituted for x. The range of f is all nonnegative reals since y is the result of an absolute value. The function is decreasing on $(-\infty, -3]$ and is increasing on $[-3, \infty)$. Domain : $(-\infty, \infty)$, Range : $[0, \infty)$; See Figure 4.

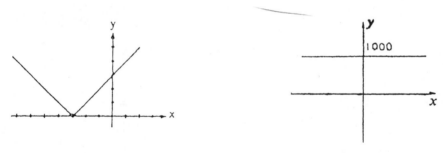

Figure 4   Figure 7

7) The domain of $f(x) = 1000$ is the set of all reals. Domain : $(-\infty, \infty)$; The range is only the number 1000. Range : $\{1000\}$; The function does not increase or decrease since it is constant over its entire domain. See Figure 7.

10) f is defined for every real number so the domain is $(-\infty, \infty)$. By examining Figure 10, we see that the range is $[0, \infty)$. The function

decreases on $(-\infty, 0]$, increases on $[0, 2]$, and is constant on $[2, \infty)$.

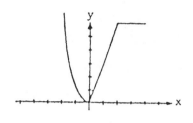

Figure 10   Figure 13

13) $f(x) = 5x^2 + 30x + 49 = 5(x^2 + 6x) + 49$

$= 5(x^2 + 6x + 9) + 49 - 45 = 5(x + 3)^2 + 4$; The vertex is the

minimum point of $(-3, 4)$. See Figure 13.

16) $(f \circ g)(x) = f(g(x)) = f\left(\dfrac{1}{x^2}\right) = \sqrt{3\left(\dfrac{1}{x^2}\right) + 2} = \sqrt{\dfrac{2x^2 + 3}{x^2}}$

$(g \circ f)(x) = g(f(x)) = g(\sqrt{3x + 2}) = \dfrac{1}{(\sqrt{3x + 2})^2} = \dfrac{1}{3x + 2}$

19) $V = \Pi r^2 h = \Pi r^3$ since $r = h$; Solving $C = 2\Pi r$ for r we have

$r = \dfrac{C}{2\Pi}$. Substituting that expression for r into the volume formula

yields $V = \Pi \left(\dfrac{C}{2\Pi}\right)^3 = \dfrac{C^3 \Pi}{8 \Pi^3} = \dfrac{C^3}{8 \Pi^2}$.

22) (a) $\dfrac{r}{x} = \dfrac{2}{4} \Rightarrow r = \dfrac{1}{2}x$

(b) The volume of water in the cone plus the volume of water in the
cup equal 5 cubic inches. $V = \frac{1}{3}\Pi r^2 h$ is the formula for the
volume of a cone and $V = \Pi r^2 h$ is the formula for the volume of
a cylinder (the cup). Adding these together and solving for y
yields : $V = \frac{1}{3}\Pi r^2 h + \Pi r^2 h \Rightarrow 5 = \frac{1}{3}\Pi \left(\frac{1}{2}x\right)^2 (x) + \Pi (2)^2 (y)$
(continued on next page)

$$\Rightarrow 5 = \frac{1}{12}\Pi x^3 + 4\Pi y \Rightarrow \frac{60}{\Pi} = x^3 + 48y \Rightarrow y = \frac{1}{48}\left(\frac{60}{\Pi} - x^3\right)$$

$$\Rightarrow y = \frac{5}{4\Pi} - \frac{x^3}{48}$$

25) The perimeter of the track is represented by the equation

P = 2x + 2Π r where x is the length of the "straightaway" and r is

the radius of the "turns". Letting P = $\frac{1}{2}$ and solving for x we obtain

$x = \frac{1 - 4\Pi r}{4}$. The area of the rectangle is then A = 2rx =

$$2r\left(\frac{1 - 4\Pi r}{4}\right) = \frac{r(1 - 4\Pi r)}{2} = -2\Pi r^2 + \frac{1}{2}r = \{\text{complete the square}\}$$

$-2\Pi\left(r - \frac{1}{8\Pi}\right)^2 + \frac{1}{32\Pi}$; Thus the dimensions that will maximize

the area of the rectangle are r = $\frac{1}{8\Pi}$ mi and x = $\frac{1}{8}$ mi.

28) (a) f (t) will be 0 when the player leaves the floor and when he

returns to the floor. Solving f (t) = 0 yields :

$0 = -\frac{1}{2}(32)t^2 + 16t \Rightarrow 0 = -16t(t - 1) \Rightarrow t = 0$ or 1. The

difference in these two times (1 second) is the *hang time*.

(b) Complete the square to find the vertex of the parabola.

f (t) = $-16t^2 + 16t = -16(t^2 - t) = -16(t^2 - t + \frac{1}{4}) + 4$

= $-16(t - \frac{1}{2})^2 + 4$; This result indicates that the player

reaches a maximum height of 4 feet at t = $\frac{1}{2}$ sec.

(c) For part (a), $0 = -\frac{1}{2}(\frac{32}{6})t^2 + 16t \Rightarrow \{\text{mult. by} -\frac{6}{8}\}$

$0 = t^2 - 6t \Rightarrow 0 = t(t - 6) \Rightarrow t = 0$ or 6 seconds.

For part (b), f (t) = $-\frac{1}{2}(\frac{32}{6})t^2 + 16t = -\frac{8}{3}t^2 + 16t =$

$-\frac{8}{3}(t^2 - 6t) = -\frac{8}{3}(t^2 - 6t + 9) + 24 = -\frac{8}{3}(t - 3)^2 + 24$.

This last form indicates that the player would reach a maximum

height of 24 feet at t = 3 seconds.

1) The graph of $f(x) = ax^3 + 2$ is a cubic polynomial with y-intercept at 2. Numbers for a that are greater than 1 in absolute value make the graph "stretch" vertically, whereas numbers for a that are less than 1 in absolute value make the graph "stretch" horizontally.

(a) $a = 2$

(b) $a = 4$

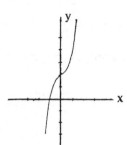

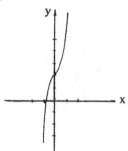

(c) $a = \frac{1}{4}$

(d) $a = -2$

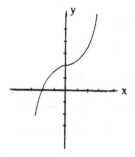

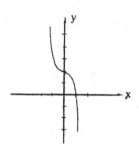

4) $f(x) = -\frac{1}{4}x^3 - 16 = -\frac{1}{4}(x^3 + 64) = -\frac{1}{4}(x + 4)(x^2 - 4x + 16)$; $f(x) = 0$ only when $x = -4$ {solving $x^2 - 4x + 16 = 0$ for x yields the square root of a negative number, indicating that $x^2 - 4x + 16$ is never equal to 0}. The y-intercept is $f(0) = -16$. $f(x) > 0$ on $(-\infty, -4)$, $f(x) < 0$ on $(-4, \infty)$; See Figure 4.

Exercises 3.1

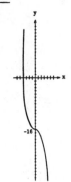

Figure 4

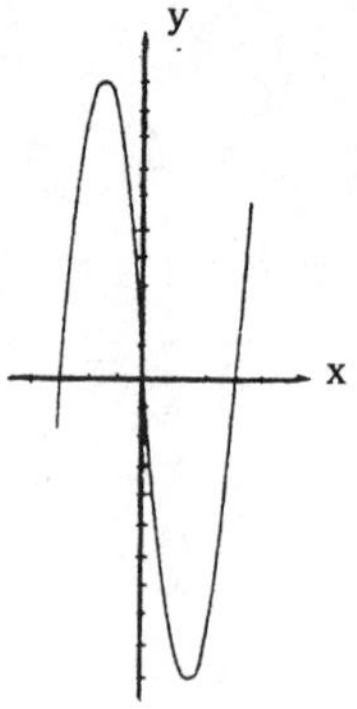

Figure 7

7) $f(x) = x^3 - 9x = x(x^2 - 9) = x(x + 3)(x - 3)$; $f(x) = 0$ when $x = \pm 3, 0$; $f(x) > 0$ on $(-3, 0)$ and $(3, \infty)$, $f(x) < 0$ on $(-\infty, -3)$ and $(0, 3)$; See Figure 7.

10) $f(x) = x^3 + x^2 - 12x = x(x^2 + x - 12) = x(x + 4)(x - 3)$; $f(x) = 0$ when $x = -4, 0, 3$; $f(x) > 0$ on $(-4, 0)$ and $(3, \infty)$, $f(x) < 0$ on $(-\infty, -4)$ and $(0, 3)$; See Figure 10.

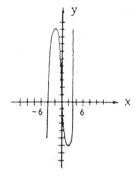

Figure 10

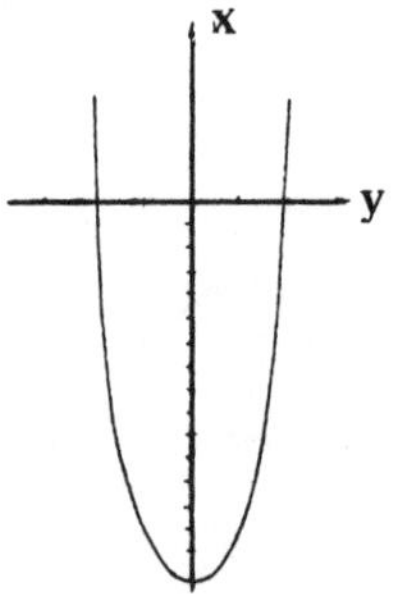

Figure 13

13) $f(x) = x^4 - 16 = (x^2 + 4)(x^2 - 4) = (x^2 + 4)(x + 2)(x - 2)$; $f(x) = 0$ when $x = \pm 2$; $f(x) > 0$ on $(-\infty, -2)$ and $(2, \infty)$, $f(x) < 0$ on $(-2, 2)$; The y-intercept is -16. See Figure 13.

16) $f(x) = x^4 - 7x^2 - 18$

$= (x^2 - 9)(x^2 + 2)$

$= (x + 3)(x - 3)(x^2 + 2);$

$f(x) = 0$ when $x = \pm 3;$

$f(x) > 0$ on $(-\infty, -3)$ and $(3, \infty)$,

$f(x) < 0$ on $(-3, 3);$ The

y-intercept is -18. See Figure 16.

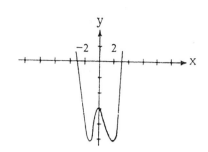

Figure 16

19) Let f be a polynomial with 0 for coefficients for all odd power terms. Since $(-x)^n = x^n$ when n is even, $f(-x) = f(x)$ and f is an even function.

22) If one zero of f is 2, then $f(2) = 0$. But $f(2) =$ $2^3 - 2(2)^2 - 16(2) + 16k = 16k - 32;$ Therefore, $16k - 32 = 0$ $\Rightarrow k = 2$. Now $f(x) = x^3 - 2x^2 - 16x + 32$ $= (x + 4)(x - 4)(x - 2);$ The other two zeros are ± 4.

25) $f(2) = 5$ and $f(3) = -5;$ By the Intermediate Value Theorem, f takes on every value between -5 and 5 in the interval $[2, 3]$ and thus takes on 0.

28) $f(3) = -60$ and $f(4) = 134;$ By the Intermediate Value Theorem, f takes on every value between -60 and 134 in the interval $[3, 4]$ and thus takes on 0.

31) $V(x) > 0$ on $(0, 10)$ and $(15, \infty)$. Allowable values for x are in the interval $(0, 10)$ since values greater than 15 would make the formula for V positive, but two of the dimensions would be negative.

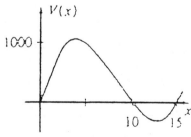

Figure 31

Exercises 3.1

34) (a) At the end of the board, $s = 10$. Letting $d = 1$ and $L = 10$ we have $1 = 100c$ (20) or $c = \frac{1}{2000}$.

(b) $d = \frac{1}{2000} (6.5)^2 (3 (10) - 6.5) \simeq 0.4964$ when $s = 6.5$;

$d = \frac{1}{2000} (6.6)^2 (3 (10) - 6.6) \simeq 0.5097$ when $s = 6.6$;

By the Intermediate Value Theorem, the function takes on every value between 0.4964 and 0.5097 in the interval [6.5, 6.6] and thus takes on $\frac{1}{2}$.

Calculator Exercises 3.1

1) We must show that the function values are of opposite sign for each pair of numbers in parts (a), (b), and (c). The Intermediate Value Theorem then assures us that 0 will be taken on by some value between the pair of numbers. The results are as follows :

(a) $f (1) = -1$ and $f (2) = 3$

(b) $f (1.5) = -0.125$ and $f (1.6) = 0.296$

(c) $f (1.53) = -0.008423$ and $f (1.54) = 0.032264$

For Exercises 4, 7, and 10, we will use a method which involves bisecting the interval. First, we evaluate the endpoints to make sure that their function values are of opposite sign. Next we evaluate the midpoint of that interval. Now use the interval with endpoints consisting of the midpoint and the endpoint with the opposite sign. Repeat this process until the desired degree of accuracy is obtained. For each problem, the work is laid out in the order that one would normally proceed and the results are rounded.

4) f (3) = -22 and f (4) = 18;   f (3.72) = -0.6589

   f (3.50) = -10.6875   f (3.73) = -0.1130

   f (3.75) =   1.0039   f (3.74) = 0.4413

   f (3.63) = -5.2072   There is a zero between 3.73

   f (3.69) = -2.2472   and 3.74.

7) f (1) = -1 and f (2) = 13;   f (1.335) = 0.0237

   f (1.500) = 1.5625   f (1.332) = 0.0033

   f (1.250) = -0.4648   f (1.330) = -0.0102

   f (1.375) = 0.3206   f (1.331) = -0.0035

   f (1.312) = -0.1269   There is a zero between

   f (1.343) = 0.0795   1.331 and 1.332.

   f (1.328) = -0.0236

10) f (-4) = -186 and f (-3) = 25;   f (-3.304) = -0.1504

   f (-3.500) = -30.6563   f (-3.300) = 0.3394

   f (-3.250) =   6.0615   f (-3.302) = 0.0951

   f (-3.375) = -9.6803   f (-3.303) = -0.0275

   f (-3.312) = -1.1447   There is a zero between

   f (-3.281) =   2.6002   -3.303 and -3.302.

   f (-3.297) =   0.7035

Exercises 3.2

1)
$$
\begin{array}{r}
x^2 + x + 2 \\
x^2 + 2x - 4 \overline{\smash{\big)}\ x^4 + 3x^3 + 0x^2 - 2x + 5} \\
\underline{x^4 + 2x^3 - 4x^2} \\
x^3 + 4x^2 \\
\underline{x^3 + 2x^2 - 4x} \\
2x^2 + 2x \\
\underline{2x^2 + 4x - 8} \\
- 2x + 13
\end{array}
$$

$q(x) = x^2 + x + 2$

$r(x) = -2x + 13$

4)
$$
\begin{array}{r}
\tfrac{3}{2}x - \tfrac{1}{2} \\
2x^3 - x + 4 \overline{\smash{\big)}\ 3x^4 - x^3 - x^2 + 3x + 4} \\
\underline{3x^4 \qquad - \tfrac{3}{2}x^2 + 6x} \\
- x^3 + \tfrac{1}{2}x^2 - 3x \\
\underline{- x^3 \qquad + \tfrac{1}{2}x - 2} \\
\tfrac{1}{2}x^2 - \tfrac{7}{2}x + 6
\end{array}
$$

$q(x) = \tfrac{3}{2}x - \tfrac{1}{2}$

$r(x) = \tfrac{1}{2}x^2 - \tfrac{7}{2}x + 6$

7)
$$
\begin{array}{r|rrrr}
2 & 2 & -3 & 4 & -5 \\
  &   & 4 & 2 & 12 \\
\hline
  & 2 & 1 & 6 & 7
\end{array}
$$

$q(x) = 2x^2 + x + 6$

$r(x) = 7$

10)
$$
\begin{array}{r|rrrr}
4 & 5 & -6 & 0 & 15 \\
  &   & 20 & 56 & 224 \\
\hline
  & 5 & 14 & 56 & 239
\end{array}
$$

$q(x) = 5x^2 + 14x + 56$

$r(x) = 239$

13)
$$
\begin{array}{r|rrrrr}
\tfrac{1}{2} & 4 & 0 & -5 & 0 & 1 \\
  &   & 2 & 1 & -2 & -1 \\
\hline
  & 4 & 2 & -4 & -2 & 0
\end{array}
$$

$q(x) = 4x^3 + 2x^2 - 4x - 2$

$r(x) = 0$

16) There are two cases; n odd and n even.

$$
\begin{array}{r|rrrrrrrr}
-1 & 1 & 0 & 0 & 0 & \cdots & 0 & 0 & 1 \\
   &   & -1 & 1 & -1 & \cdots & -1 & 1 & -1 \\
\hline
   & 1 & -1 & 1 & -1 & \cdots & -1 & 1 & 0
\end{array}
\quad \text{(n odd)}
$$

$q(x) = x^{n-1} - x^{n-2} + x^{n-3} - x^{n-4} + \cdots - x + 1,\ r(x) = 0$

$$
\begin{array}{r|rrrrrrrr}
-1 & 1 & 0 & 0 & 0 & \cdots & 0 & 0 & 1 \\
   &   & -1 & 1 & -1 & \cdots & 1 & -1 & 1 \\
\hline
   & 1 & -1 & 1 & -1 & \cdots & 1 & -1 & 2
\end{array}
\quad \text{(n even)}
$$

$q(x) = x^{n-1} - x^{n-2} + x^{n-3} - x^{n-4} + \cdots + x - 1,\ r(x) = 2$

19) Synthetically dividing $f(x)$ by $x + 2$ yields a remainder of -23. By the Remainder Theorem, $f(-2) = -23$.

22) Synthetically dividing $f(x)$ by $x + 1$ yields a remainder of -6. By the Remainder Theorem, $f(-1) = -6$.

25) Synthetically dividing $f(x)$ by $x - 4$ yields a remainder of 3277. By the Remainder Theorem, $f(4) = 3277$.

28) Synthetically dividing $f(x)$ by $x - (1 + \sqrt{2})$ yields a remainder of $-10 - \sqrt{2}$. By the Remainder Theorem, $f(1 + \sqrt{2}) = -10 - \sqrt{2}$.

31)
$$
\begin{array}{r|rrrr}
\frac{1}{2} & 4 & -6 & 8 & -3 \\
   &   & 2 & -2 & 3 \\
\hline
   & 4 & -4 & 6 & 0
\end{array}
$$
$q(x) = 4x^2 - 4x + 6$

$r(x) = 0 \implies \frac{1}{2}$ is a zero of $f(x)$

34)
$$
\begin{array}{r|rrrr}
1 & k^2 & 0 & -4k & -3 \\
  &   & k^2 & k^2 & k^2 - 4k \\
\hline
  & k^2 & k^2 & k^2 - 4k & k^2 - 4k - 3
\end{array}
$$

We find that the remainder is $k^2 - 4k - 3$. For $f(x)$ to be divisible by $x - 1$, the remainder must be 0. $k^2 - 4k - 3 = 0 \implies k = \dfrac{4 \pm \sqrt{16 + 12}}{2} = 2 \pm \sqrt{7}$

Exercises 3.2

37) $f(c) = 3c^4 + c^2 + 5$; This value is greater than or equal to 5 when any real number is substituted for c. Thus, f could never have a remainder of 0 and therefore no factor of the form $x - c$.

40) If $f(x) = x^n + y^n$ and n is odd, then $f(-y) = (-y)^n + y^n = -y^n + y^n = 0$. Thus, $x + y$ is a factor of f.

Exercises 3.3

1) $(3 + 2i) + (-5 + 4i) = (3 + -5) + (2 + 4)i = -2 + 6i$

4) $(5 + 7i) + (-8 - 4i) = (5 + -8) + (7 + -4)i = -3 + 3i$

7) $7 - (3 - 7i) = (7 - 3) - (-7)i = 4 + 7i$

10) $(10 + 7i) - 12i = 10 + (7 - 12)i = 10 - 5i$

13) $(-7 + i)(-3 + i) = [(-7)(-3) + (1)(1) i^2] + [(-7)(1) + (-3)(1)] i$
$= [21 + -1] + [-7 + -3] i = 20 - 10i$

16) $7i(13 + 8i) = (7)(8) i^2 + (7)(13) i = -56 + 91i$

19) $(3 + i)^2 (3 - i)^2 = [(3 + i)(3 - i)]^2 = [(3)^2 - (i)^2]^2 = [9 + 1]^2 = (10)^2 = 100$ {recognize the above as the difference of squares}

22) $i^{25} = i^{24} i^1$ {separate into one exponent that is a multiple of 4 since $i^4 = 1$} $= (i^4)^6 i = (1)^6 i = 1 i = i$

25) $\dfrac{7}{5 - 6i} = \dfrac{7}{5 - 6i} \cdot \dfrac{5 + 6i}{5 + 6i} = \dfrac{7(5 + 6i)}{(5)^2 - (6i)^2} = \dfrac{7(5 + 6i)}{25 - (-36)} = \dfrac{35 + 42i}{61} =$
$\dfrac{35}{61} + \dfrac{42}{61} i$

28) $\dfrac{4 + 3i}{-1 + 2i} = \dfrac{4 + 3i}{-1 + 2i} \cdot \dfrac{-1 - 2i}{-1 - 2i} = \dfrac{(-4 + 6) + (-3 + -8) i}{(-1)^2 - (2i)^2} = \dfrac{2 - 11i}{1 - (-4)} =$
$\dfrac{2}{5} - \dfrac{11}{5} i$

31) $\dfrac{21 - 7i}{i} = \dfrac{21 - 7i}{i} \cdot \dfrac{-i}{-i} = \dfrac{7i^2 - 21i}{1} = -7 - 21i$

{in Exercise 31, select $-i$ since $0 - i$ is the conjugate of $0 + i$}

34) For the two sides to equal each other, we must have the "real parts" equal and the "imaginary parts" equal. If $2x = 8$, then $x = 4$. Now $3x + y = -4$ or $y = -4 - 3x = -4 - 3(4) = -16$.

37) $\{a = 1, b = 2, c = 5\}$ $x = \dfrac{-2 \pm \sqrt{4 - 20}}{2} = \dfrac{-2 \pm 4i}{2} = -1 \pm 2i$

40) $\{a = -3, b = 1, c -5\}$ $x = \dfrac{-1 \pm \sqrt{1 - 60}}{-6} = \dfrac{1}{6} \pm \dfrac{\sqrt{59}}{6} i$

43) $x^6 - 64 = (x^3 + 8)(x^3 - 8)$

$\qquad = (x + 2)(x^2 - 2x + 4)(x - 2)(x^2 + 2x + 4)$

$x = \dfrac{2 \pm \sqrt{4 - 16}}{2} = 1 \pm \sqrt{3}\, i$ and -2 for $x^3 + 8$

$x = \dfrac{-2 \pm \sqrt{4 - 16}}{2} = -1 \pm \sqrt{3}\, i$ and 2 for $x^3 - 8$

46) $0 = 27x^4 + 21x^2 + 4 = (3x^2 + 1)(9x^2 + 4)$

$3x^2 + 1 = 0 \implies 3x^2 = -1 \implies x^2 = -\dfrac{1}{3} \implies x = \pm\sqrt{-\dfrac{1}{3}} = \pm\dfrac{\sqrt{3}}{3} i$

$9x^2 + 4 = 0 \implies 9x^2 = -4 \implies x^2 = -\dfrac{4}{9} \implies x = \pm\sqrt{-\dfrac{4}{9}} = \pm\dfrac{2}{3} i$

49) Let $z = a + bi$ and $w = c + di$. (LHS denotes Left Hand Side and RHS denotes Right Hand Side)

(a) LHS $= \overline{z + w}$

$\qquad = \overline{(a + bi) + (c + di)}$

$\qquad = \overline{(a + c) + (b + d)i}$

$\qquad = (a + c) - (b + d)i$

$\qquad = (a - bi) + (c - di)$

$\qquad = \overline{a + bi} + \overline{c + di}$

$\qquad = \bar{z} + \bar{w} =$ RHS

<u>Exercises 3.3</u>

(b) LHS $= \overline{z \cdot w}$

$\qquad = \overline{(a + bi) \cdot (c + di)}$

$\qquad = \overline{(ac - bd) + (ad + bc)i}$

$\qquad = (ac - bd) - (ad + bc)i$

$\qquad = ac - adi - bd - bci$

$\qquad = a(c - di) - bi(c - di)$

$\qquad = (a - bi) \cdot (c - di)$

$\qquad = \overline{a + bi} \cdot \overline{c + di}$

$\qquad = \overline{z} \cdot \overline{w} = $ RHS

(c) LHS $= \overline{z^2}$

$\qquad = \overline{(a + bi)^2}$

$\qquad = \overline{a^2 + 2abi - b^2}$

$\qquad = \overline{(a^2 - b^2) + 2abi}$

$\qquad = (a^2 - b^2) - 2abi$

$\qquad = a^2 - 2abi - b^2$

$\qquad = (a - bi)^2$

$\qquad = (\overline{z})^2 = $ RHS

LHS $= \overline{z^3}$

$\qquad = \overline{(a + bi)^3}$

$\qquad = \overline{a^3 + 3a^2bi - 3ab^2 - b^3i}$

$\qquad = \overline{(a^3 - 3ab^2) + (3a^2b - b^3)i}$

$\qquad = a^3 - 3ab^2 - 3a^2bi + b^3i$

$\qquad = a^3 - 3a^2bi - 3ab^2 + b^3i$

$\qquad = (a - bi)^3$

$\qquad = (\overline{z})^3 = $ RHS

$$\text{LHS} = \overline{z^4}$$

$$= \overline{(a + bi)^4}$$

$$= \overline{a^4 + 4a^3bi - 6a^2b^2 - 4ab^3i + b^4}$$

$$= \overline{(a^4 - 6a^2b^2 + b^4) + (4a^3b - 4ab^3)i}$$

$$= a^4 - 6a^2b^2 + b^4 - 4a^3bi + 4ab^3i$$

$$= a^4 - 4a^3bi - 6a^2b^2 + 4ab^3i + b^4$$

$$= (a - bi)^4$$

$$= (\bar{z})^4 = \text{RHS}$$

(d) Let $z = a + bi$.

1) Suppose $\bar{z} = z$. Then $a - bi = a + bi$

$$\Rightarrow \quad -bi = bi$$

$$\Rightarrow \quad 0 = 2bi$$

$$\Rightarrow \quad b = 0$$

$$\Rightarrow z = a \text{ which is real.}$$

2) Suppose now that z is real. Then $z = a + 0i = a$ and $\bar{z} = a - 0i = a$. So $\bar{z} = z$.

By 1) and 2) we conclude $\bar{z} = z \iff z$ is real.

1) Since 5, −2, and −3 are zeros, $(x - 5)$, $(x + 2)$, and $(x + 3)$ are factors; Thus, $f(x) = a(x - 5)(x + 2)(x + 3)$ where a is to be determined; $f(2) = a(2 - 5)(2 + 2)(2 + 3) = a(-3)(4)(5) = -60a$; But $f(2) = 4$, so $-60a = 4$ which implies that $a = -\frac{1}{15}$;
$f(x) = -\frac{1}{15}(x - 5)(x + 2)(x + 3) = -\frac{1}{15}x^3 + \frac{19}{15}x + 2$

4) $f(x) = a(x - \sqrt{2})(x - \Pi)(x)$; $f(0) = a(-\sqrt{2})(-\Pi)(0) = 0$

$\Rightarrow$ a is any nonzero number. If $a = 1$, then

$f(x) = x^3 - (\Pi + \sqrt{2})x^2 + \Pi\sqrt{2}x$

Exercises 3.4

7) $f(x) = (x + 2)^2 (x - 3)^2 = x^4 - 2x^3 - 11x^2 + 12x + 36$ or any nonzero multiple of f.

10) $f(x) = a(x - 1)^2 (x + 1)^2 (x)^3$; $f(2) = a(1)(9)(8) = 72a = 36 \Rightarrow$
    $a = \frac{1}{2}$ so $f(x) = \frac{1}{2}x^7 - x^5 + \frac{1}{2}x^3$

13) $f(x) = 2x^5 - 8x^4 - 10x^3 = 2x^3(x^2 - 4x - 5) = 2x^3(x - 5)(x + 1)$;
    Thus, 0 is a root of multiplicity 3; 5 and −1 are roots of multiplicity 1.

16) $f(x) = (2x^2 + 13x - 7)^3 = [(2x - 1)(x + 7)]^3 = (2x - 1)^3(x + 7)^3$;
    Thus $\frac{1}{2}$ and −7 are roots each having multiplicity 3.

19) Since we know that −3 is a zero of multiplicity 2, we can synthetically divide −3 twice and be left with the polynomial $x^2 + x - 2$. This factors into $(x + 2)(x - 1)$. The complete factorization of $f(x)$ is $(x + 3)^2(x + 2)(x - 1)$.

22) As in Exercise 19, we use synthetic division 4 times with −1 as our divisor to reduce $f(x)$ to $x - 3$. Thus, $f(x) = (x + 1)^4(x - 3)$.

In Exercises 25 & 28 the types of possible solutions are listed in the following order : positive, negative, nonreal complex.

25) The number of sign changes in $f(x)$ is 0.
    The number of sign changes in $f(-x) = -4x^3 + 2x^2 + 1$ is 1.
    Can only be 0, 1, 2

28) The number of sign changes in $f(x)$ is 4.
    The number of sign changes in $f(-x) = 2x^4 + x^3 + x^2 + 3x + 4$ is 0.
    Either 4, 0, 0; 2, 0, 2; or 0, 0, 4

31) Synthetically dividing 5 into the polynomial yields a bottom row consisting of all nonnegative numbers. Synthetically dividing −2 into the polynomial yields a bottom row that alternates in sign.

These results indicate that the upper bound is 5 and the lower bound is -2.

34) Synthetically dividing 5 into the polynomial yields a bottom row consisting of all nonnegative numbers. Synthetically dividing -1 into the polynomial yields a bottom row that alternates in sign. These results indicate that the upper bound is 5 and the lower bound is -1.

37) Follow the hint in the text. Assume the a_j's and the b_j's are not the same and hence $f(x) \neq g(x)$. There are more than n distinct values that make f and g equal and these values will make $h(x)$ zero. But now $h(x)$ has more than n distinct zeros and is only an n degree polynomial. Since h can have at most n zeros we conclude the assumption that $f(x) \neq g(x)$ is incorrect and that $a_j = b_j$ for all j.

It is helpful to remember that if $a + bi$ is a root, then $x^2 - (2a)x + (a^2 + b^2)$ is the associated quadratic factor for Exercises 1, 4, and 7.

1) Since $4 + i$ is a root, so is $4 - i$. The polynomial has the form
$$(x - (4 + i))(x - (4 - i)) = x^2 - 8x + 17.$$

4) Since $1 + 5i$ is a root, so is $1 - 5i$. The polynomial has the form
$$(x - (1 + 5i))(x - (1 - 5i))(x + 2) = (x^2 - 2x + 26)(x + 2) = x^3 + 22x + 52.$$

7) Since $5i$ and $1 + i$ are roots, so are $-5i$ and $1 - i$. The polynomial has the form $(x - (5i))(x + (5i))(x - (1 + i))(x - (1 - i))(x) =$
$(x^2 + 25)(x^2 - 2x + 2)(x) = x^5 - 2x^4 + 27x^3 - 50x^2 + 50x$

Exercises 3.5

10) The theorem applies only to polynomials with real coefficients, whereas the polynomial in question has complex coefficients.

For Exercises 13, 16, 19, 22, 25, and 28, the answers are listed in the following form : (a) the choices for c, (b) the choices for d, (c) the possible rational roots (c/d), and (d) the actual roots of the equation. Use synthetic division to determine that the listed values are indeed the solutions.

13) (a) 1, 2, 3, 5, 6, 10, 15, 30   (b) $\pm 1, \pm 2$

    (c) $\pm 1, \pm 2, \pm 3, \pm 5, \pm 6, \pm 10, \pm 15, \pm 30, \pm \frac{1}{2}, \pm \frac{3}{2}, \pm \frac{5}{2}, \pm \frac{15}{2}$

    (d) $-3, 2, \frac{5}{2}$

16) (a) 1, 2, 3, 6   (b) $\pm 1, \pm 3$

    (c) $\pm 1, \pm 2, \pm 3, \pm 6, \pm \frac{1}{3}, \pm \frac{2}{3}$

    (d) -1 (multiplicity 2), $\frac{1}{3}$, 2, 3

19) (a) 1, 2, 3, 6   (b) $\pm 1, \pm 2, \pm 3, \pm 6$

    (c) $\pm 1, \pm 2, \pm 3, \pm 6, \pm \frac{1}{2}, \pm \frac{1}{3}, \pm \frac{1}{6}, \pm \frac{2}{3}, \pm \frac{3}{2}$

    (d) $-3, -\frac{2}{3}, \frac{1}{2}$

22) (a) 1, 2, 4, 5, 10, 20   (b) $\pm 1, \pm 3$

    (c) $\pm 1, \pm 2, \pm 4, \pm 5, \pm 10, \pm 20, \pm \frac{1}{3}, \pm \frac{2}{3}, \pm \frac{4}{3}, \pm \frac{5}{3}, \pm \frac{10}{3}, \pm \frac{20}{3}$

    (d) $\frac{4}{3}$ is the only rational root, the other two can be found using the

      Quadratic Formula. They are $\dfrac{-1 \pm \sqrt{19}\, i}{2}$.

25) (a) 1, 2, 4, 8, 16   (b) $\pm 1, \pm 3, \pm 9$

    (c) $\pm 1, \pm 2, \pm 4, \pm 8, \pm 16, \pm \frac{1}{3}, \pm \frac{2}{3}, \pm \frac{4}{3}, \pm \frac{8}{3}, \pm \frac{16}{3}, \pm \frac{1}{9}, \pm \frac{2}{9}, \pm \frac{4}{9}, \pm \frac{8}{9},$

      $\pm \frac{16}{9}$ (d) $-2, 1, \frac{2}{3}, -\frac{4}{3}$

28) (a) 1, 2, 3, 5, 6, 9, 10, 15, 18, 27, 30, 45, 54, 90, 135, 270

    (b) $\pm 1, \pm 2, \pm 4$

    (c) $\pm 1, \pm 2, \pm 3, \pm 5, \pm 6, \pm 9, \pm 10, \pm 15, \pm 18, \pm 27, \pm 30, \pm 45,$

$\pm 54,\ \pm 90,\ \pm 135,\ \pm 270,\ \pm\frac{1}{2},\ \pm\frac{1}{4},\ \pm\frac{3}{2},\ \pm\frac{3}{4},\ \pm\frac{5}{2},\ \pm\frac{5}{4},\ \pm\frac{9}{2},\ \pm\frac{9}{4},$

$\pm\frac{15}{2},\ \pm\frac{15}{4},\ \pm\frac{27}{2},\ \pm\frac{27}{4},\ \pm\frac{45}{2},\ \pm\frac{45}{4},\ \pm\frac{135}{2},\ \pm\frac{135}{4}$

(d) $-5,\ -3,\ \pm\frac{3}{2},\ 2$

For Exercises 31 and 34, it must be shown that none of the possible rational roots listed satisfy the equation.

31) $\frac{c}{d} = \pm 1,\ \pm 2$

34) $\frac{c}{d} = \pm 1,\ \pm\frac{1}{2},\ \pm 5,\ \pm\frac{5}{2}$

37) Since n is odd and nonreal complex zeros occur in conjugate pairs for polynomials with real coefficients, there must be at least one real zero.

40) By the Theorem on Rational Zeros, r is of the form $\frac{c}{d}$ where c is a factor of a_0 and d is a factor of a_n. Since a_n is 1, its divisors are ± 1. Thus, r will be an integer and a factor of a_0.

43) $T = 0.05t\,(t - 12)\,(t - 24)$ and $T = 32$ yield the equations :
$32 = 0.05t\,(t - 12)\,(t - 24)\ \leftrightarrow\ 640 = t\,(t^2 - 36t + 288)\ \leftrightarrow$
$t^3 - 36t^2 + 288t - 640 = 0$. Use the Theorem on Rational Zeros to find that one solution is $t = 4$. The other two are found by using the Quadratic Formula. They are $16 \pm 4\sqrt{6}$. $\{25.8$ and $6.2\}$ Since $0 \le t \le 24$, $t = 4$ (10:00 A.M.) and $t = 16 - 4\sqrt{6}$ (12:12 P.M.) are the times when the temperature was 32°F.

46) $V(x) = 10\Pi x^2 + \frac{4}{3}\Pi x^3$ and $V(x) = 27\Pi$ yield the equations :
$27\Pi = \frac{4}{3}\Pi x^3 + 10\Pi x^2\ \leftrightarrow\ 4x^3 + 30x^2 - 81 = 0\ \leftrightarrow$
$(2x - 3)\,(2x^2 + 18x + 27) = 0$; The solutions to the quadratic are
$x = \dfrac{-18 \pm \sqrt{324 - 216}}{4} = \dfrac{-18 \pm 6\sqrt{3}}{4} \simeq -1.9,\ -7.1$; Neither of
these are possible answers, so the radius is 1.5 feet. $\{$from $2x - 3\}$

Exercises 3.6

For Exercises 1, 4, 7, 10, 13, 16, and 19, the information about the graphs is listed in the following order : (a) the x-intercept(s), (b) the equation(s) of the vertical asymptote(s), (c) the y-intercept, and (d) the equation of the horizontal asymptote.

1) (a) None   (b) x = -2   (c) $\frac{1}{2}$ (d) y = 0;   See Figure 1.

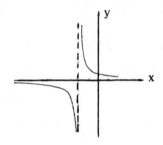

Figure 1

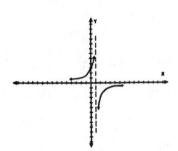

Figure 4

4) (a) None   (b) x = 1   (c) 3   (d) y = 0;   See Figure 4.

7) (a) None   (b) x = 1   (c) 4   (d) y = 0;   See Figure 7.

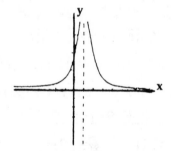

Figure 7

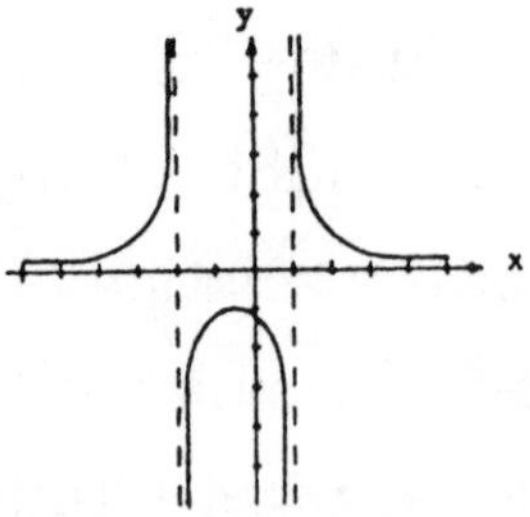

Figure 10

10) (a) None   (b) x = -2, x = 1   (c) -1   (d) y = 0;   See Figure 10.

13) (a) 0   (b) x = 2, x = 5   (c) 0   (d) y = 1;   See Figure 13.

16) (a) ±2   (b) x = 0   (c) None   (d) y = 1;   See Figure 16.

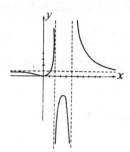

Figure 13

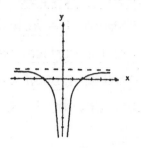

Figure 16

19) (a) None   (b) x = −3, x = 0, x = 2   (c) None   (d) y = 0;   See Figure 19.

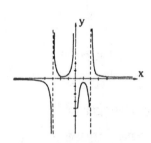

Figure 19

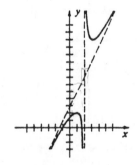

Figure 22

22) x-intercepts : −1, $\frac{3}{2}$;   y-intercept : $\frac{3}{2}$;   vertical asymptote : x = 2; oblique asymptote : y = 2x + 3;   See Figure 22.

25) (a) The radius of the outside cylinder is (r + 0.5) feet and its height is (h + 1) feet. Since the volume is 16Π ft³, we have

$$16\Pi = \Pi\,(r + 0.5)^2\,(h + 1) \;\Rightarrow\; h = \frac{16}{(r + 0.5)^2} - 1$$

(b) $V\,(r) = \Pi r^2 h = \Pi r^2 \left[\dfrac{16}{(r + 0.5)^2} - 1 \right]$

(c) r and h must both be positive. $h > 0 \;\Rightarrow\; \dfrac{16}{(r + 0.5)^2} - 1 > 0$

$\Rightarrow\; 16 > (r + 0.5)^2 \;\Rightarrow\; |r + 0.5| < 4 \;\Rightarrow\; -4.5 < r < 3.5$

(continued on next page)

Exercises 3.6

The last inequality combined with $r > 0$ means that the excluded values are $r \leq 0$ and $r \geq 3.5$.

28) As $S \rightarrow \infty$, $R \rightarrow a$. The interpretation of the constant a is that it is the limiting value of recruitment, or, as it states in the last line of Exercise 28, the recruitment is more or less constant.

Exercises 3.7

1) x-intercepts : -2, 1, 3; y-intercept : $\frac{3}{2}$; See Figure 1.

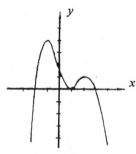

Figure 1

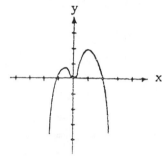

Figure 4

4) x-intercepts : -1, 0, 2; y-intercept : 0; See Figure 4.

7) no x-intercepts; y-intercept : -2; vertical asymptote : $x = -1$; horizontal asymptote : $y = 0$; See Figure 7.

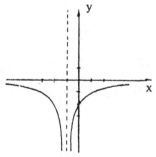

Figure 7

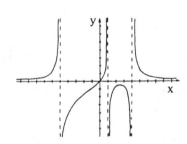

Figure 10

10) x-intercept : 0; y-intercept : 0; vertical asymptotes : $x = -5$, $x = 1$, and $x = 4$; horizontal asymptote : $y = 0$; See Figure 10.

13) Using long division, we obtain the following results :

$q(x) = 3x^2 + 2$, $r(x) = -21x^2 + 5x - 9$

16) Using long division, we obtain the following results :

$q(x) = 4x - 1$, $r(x) = 2x - 1$

19)

$$
\begin{array}{r|rrrrrr}
-2 & 6 & 0 & 0 & -4 & 0 & 8 \\
   &   & -12 & 24 & -48 & 104 & -208 \\
\hline
   & 6 & -12 & 24 & -52 & 104 & -200
\end{array}
$$

$q(x) = 6x^4 - 12x^3 + 24x^2 - 52x + 104$, $r(x) = -200$

22) $f(x) = a(x - (1 - i))(x - (1 + i))(x - 3)(x)$;

$f(2) = a(1 + i)(1 - i)(-1)(2) = -4a = -1 \Rightarrow a = \frac{1}{4}$; Therefore, f

has the form $f(x) = \frac{1}{4}(x - (1 - i))(x - (1 + i))(x - 3)(x) =$

$\frac{1}{4}(x^2 - 2x + 2)(x - 3)(x) = \frac{1}{4}x(x^3 - 5x^2 + 8x - 6) =$

$\frac{1}{4}x^4 - \frac{5}{4}x^3 + 2x^2 - \frac{3}{2}x$

25) $(x^2 - 2x + 1)^2(x^2 + 2x - 3) = ((x - 1)^2)^2(x + 3)(x - 1) =$

$(x - 1)^5(x + 3)$; 1 is a root of multiplicity 5 and -3 is a root of

multiplicity 1.

28) (a) $f(x) = x^5 - 4x^3 + 6x^2 + x + 4$, $f(-x) = -x^5 + 4x^3 + 6x^2 - x + 4$;

There are 2 sign changes in $f(x)$ and 3 sign changes in $f(-x)$.

There are either 2 positive and 3 negative solutions or 2

2 positive, 1 negative, and 2 nonreal complex solutions or 3

negative and 2 nonreal complex solutions or 1 negative and 4

nonreal complex solutions.

(b) Upper bound is 2, lower bound is -3

31) Choices for c are 1 and 3; choices for d : $\pm 1, \pm 2, \pm 4, \pm 8, \pm 16$;

possible rational roots : $\pm 1, \pm 3, \pm \frac{1}{2}, \pm \frac{3}{2}, \pm \frac{1}{4}, \pm \frac{3}{4}, \pm \frac{1}{8}, \pm \frac{3}{8}, \pm \frac{1}{16},$

$\pm \frac{3}{16}$; actual solutions : $-\frac{1}{2}, \frac{1}{4}, \frac{3}{2}$

1) y-intercept : 1; horizontal asymptote : y = 0;   See Figure 1.

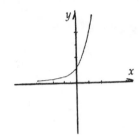

Figure 1   Figure 4

4) y-intercept : 1; horizontal asymptote : y = 0;   See Figure 4.

7) y-intercept : −1; horizontal asymptote : y = 0;   See Figure 7.

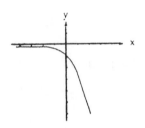

Figure 7   Figure 10

10) y-intercept : 3; horizontal asymptote : y = 2;   See Figure 10.

13) y-intercept : 1; horizontal asymptote : y = 0;   See Figure 13.

16) y-intercept : 1; horizontal asymptote : y = 0;   See Figure 16.

19) y-intercept : 8; horizontal asymptote : y = 0; This graph could be
    thought of as a shift of $y = 2^{-x}$ to the right 2 units. See Figure 19.

22) y-intercept : $\frac{1}{2}$; horizontal asymptote : y = 0; The maximum point
    to the left of the y-axis is (−1, 1). See Figure 22.

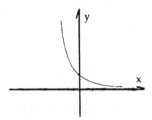

Figure 13

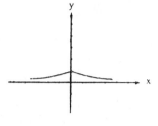

Figure 16

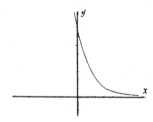

Figure 19

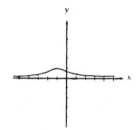

Figure 22

25) (a) $N(1) = 100(0.9)^1 = 90$

    (b) $N(5) = 100(0.9)^5 \simeq 59$

    (c) $N(10) = 100(0.9)^{10} \simeq 35$

28) (a) 2:00 P.M. corresponds to

    $t = 1$ and $f(1) = 87.5°$;

    3:30 P.M. corresponds to

    $t = \frac{5}{2}$ and $f(\frac{5}{2}) \simeq 76.6°$;

    4:00 P.M. corresponds to

    $t = 3$ and $f(3) \simeq 75.8°$

    (b) When $t = 0$, $f(t) = 125°$.

    As $t \to \infty$, $f(t) \to 75°$.

    See Figure 28.

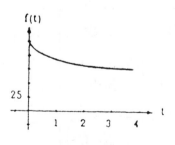

Figure 28

Exercises 4.1

31) When $t = 1600$, $q(t) = \frac{1}{2}q_0$. Thus $\frac{1}{2}q_0 = q_0 \cdot 2^{1600k}$ $\Rightarrow$

$2^{-1} = 2^{1600k}$ $\Rightarrow$ $1600k = -1$ $\Rightarrow$ $k = -\frac{1}{1600}$.

34) Using $A = P\left(1 + \frac{r}{n}\right)^{nt}$, we have $P = 5000$, $r = 0.10$, $n = 2$,

and $t = 1$. Substituting those values into the formula yields :

$5000 = P\left(1 + \frac{0.10}{2}\right)^{2 \cdot 1}$ $\Rightarrow$ $P = \frac{5000}{(1.05)^2} \approx \$4{,}535.15$

37) Because a^x is not always a real number if $a < 0$. As an example,

$(-25)^{1/2} = \sqrt{-25} = 5i.$

40) If $a > 1$, $y = a^x$ is an increasing exponential function and $y = a^{-x}$ is
the decreasing exponential function that can be obtained by reflecting
$y = a^x$ about the y-axis.

Calculator Exercises 4.1

1) The suggested points are listed in
the following table : See Figure 1.

x	$(1.8)^x$	x	$(1.8)^x$
-3.0	0.1715	0.5	1.3416
-2.5	0.2300	1.0	1.8000
-2.0	0.3086	1.5	2.4150
-1.5	0.4141	2.0	3.2400
-1.0	0.5556	2.5	4.3469
-0.5	0.7454	3.0	5.8320
0.0	1.0000		

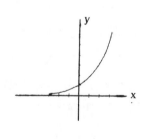

Figure 1

4) $A(t) = 1000\left[1 + \dfrac{0.06}{12}\right]^{12t} = 1000\,(1.005)^{12t}$

   (a) $A(1) \simeq \$1{,}061.68$ (b) $A(2) \simeq \$1{,}127.16$

   (c) $A(5) \simeq \$1{,}348.85$ (d) $A(10) \simeq \$1{,}819.40$

7) $A = 500\left[1 + \dfrac{0.06}{52}\right]^{(52)\,\cdot\,(1)} \simeq \$530.90$

1) (a) y-intercept : 1; horizontal asymptote : $y = 0$;   See Figure 1 (a).

   (b) y-intercept : -1; horizontal asymptote : $y = 0$;   See Figure 1 (b).

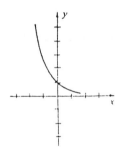

         Figure 1 (a)   Figure 1 (b)

4) (a) y-intercept : 1; horizontal asymptote : $y = 0$;   See Figure 4 (a).

   (b) y-intercept : -2; horizontal asymptote : $y = 0$;   See Figure 4 (b).

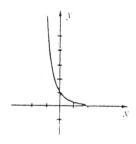
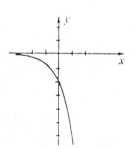

         Figure 4 (a)   Figure 4 (b)

Exercises 4.2

7) y-intercept : 1;

   horizontal asymptote : $y = 0$;

   Notice that f is an even

   function. See Figure 7.

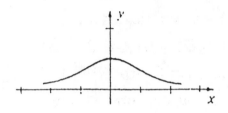

Figure 7

10) $f(x) = -x^2 e^{-x} + 2xe^{-x} = xe^{-x}(-x + 2)$; e^{-x} can never equal 0 so the

   zeros of f are 0 and 2. {from the factors x and $(-x + 2)$}

13) $\dfrac{(e^x + e^{-x})(e^x + e^{-x}) - (e^x - e^{-x})(e^x - e^{-x})}{(e^x + e^{-x})^2} = $ {Note that $e^x e^{-x} = e^{x-x} = e^0 = 1$}

$$\frac{(e^{2x} + 2 + e^{-2x}) - (e^{2x} - 2 + e^{-2x})}{(e^x + e^{-x})^2} = \frac{4}{(e^x + e^{-x})^2}$$

Calculator Exercises 4.2

1) $W(30) = 68e^{(0.2)(30)} \simeq 27{,}433$ mg or 27.4 g

4) 2000 corresponds to $t = 20$; $N(20) = 651e^{0.4} \simeq 971.2$ million

7) (a) 1990 corresponds to $t = 12$; $N(12) = 5000e^{0.564} \simeq 8{,}788$

   (b) 2000 corresponds to $t = 22$; $N(22) = 5000e^{1.034} \simeq 14{,}061$

10) Consider A as a function of t with $c = 50$.

   i.e. $A(t) = 50e^{-0.00495t}$

   (a) $A(30) = 50e^{-0.1485} \simeq 43.10$ mg

   (b) $A(180) = 50e^{-0.891} \simeq 20.51$ mg

   (c) $A(365) = 50e^{-1.80675} \simeq 8.21$ mg

1) $4^3 = 64 \iff \log_4 64 = 3$

4) $5^3 = 125 \iff \log_5 125 = 3$

7) $t^r = s \iff \log_t s = r$

10) $\log_3 81 = 4 \iff 3^4 = 81$

13) $\log_7 1 = 0 \iff 7^0 = 1$

16) $\log_v w = q \iff v^q = w$

19) $\log_{10} 100 = \log_{10} 10^2 = 2$

22) $\log_{10} 0.0001 = \log_{10} 10^{-4} = -4$

25) $\log_9 x = \frac{3}{2} \implies x = 9^{3/2} = (\sqrt{9})^3 = 3^3 = 27$

28) $\log_{10} x^2 = -4 \implies x^2 = 10^{-4} = \frac{1}{10000} \implies x = \pm \frac{1}{100}$

31) $\log_2 x - \log_2 (x + 1) = 3 \log_2 4 \implies \log_2 \left(\frac{x}{x+1} \right) = \log_2 4^3 \implies$

$\frac{x}{x + 1} = 64 \implies x = 64x + 64 \implies x = -\frac{63}{64}$; since this value

would make the argument of $\log_2 x$ negative, there is <u>no solution</u>.

34) $\frac{1}{2} \log_5 (x - 2) = 3 \log_5 2 - \frac{3}{2} \log_5 (x - 2) \implies 2 \log_5 (x - 2) = \log_5 2^3$

$\implies \log_5 (x - 2)^2 = \log_5 8 \implies x^2 - 4x + 4 = 8 \implies x^2 - 4x - 4 = 0$

$\implies x = \frac{4 \pm \sqrt{16 + 16}}{2} = 2 \pm 2\sqrt{2}$; $2 - 2\sqrt{2}$ is extraneous since it

would make the argument of $\log_5 (x - 2)$ negative, so the solution is

$x = 2 + 2\sqrt{2}$

37) $\log_a \frac{\sqrt{x}\, z^2}{y^4} = \log_a (\sqrt{x}\, z^2) - \log_a y^4 = \log_a \sqrt{x} + \log_a z^2 - \log_a y^4 =$

$\frac{1}{2} \log_a x + 2 \log_a z - 4 \log_a y$

40) $\log_a \frac{\sqrt{x}\, y^6}{\sqrt[3]{z^2}} = \log_a (\sqrt{x}\, y^6) - \log_a (\sqrt[3]{z^2}) =$

$\log_a \sqrt{x} + \log_a y^6 - \log_a \sqrt[3]{z^2} = \frac{1}{2} \log_a x + 6 \log_a y - \frac{2}{3} \log_a z$

Exercises 4.3

43) $2 \log_a x + \frac{1}{3} \log_a (x - 2) - 5 \log_a (2x + 3)$

$= \log_a x^2 + \log_a (x - 2)^{1/3} - \log_a (2x + 3)^5$

$= \log_a x^2 (x - 2)^{1/3} - \log_a (2x + 3)^5 = \log_a \left[\dfrac{x^2 \sqrt[3]{x - 2}}{(2x + 3)^5} \right]$

46) $2 \log_a \dfrac{y^3}{x} - 3 \log_a y + \frac{1}{2} \log_a x^4 y^2$

$= \log_a \left(\dfrac{y^3}{x} \right)^2 + \log_a (x^4 y^2)^{\frac{1}{2}} - \log_a y^3 = \log_a \dfrac{y^6}{x^2} + \log_a x^2 y - \log_a y^3$

$= \log_a \left(\dfrac{y^6 x^2 y}{x^2} \right) - \log_a y^3 = \log_a \left(\dfrac{y^7}{y^3} \right) = \log_a y^4$

49) $N = 10^4 (3)^t \implies \dfrac{N}{10^4} = 3^t \implies t = \log_3 \left(\dfrac{N}{10^4} \right)$

Exercises 4.4

1) x-intercept : 1; vertical asymptote : x = 0; See Figure 1.

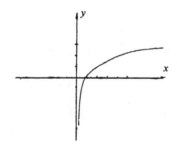

Figure 1

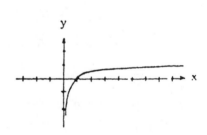

Figure 4

4) x-intercept : 1; vertical asymptote : x = 0; See Figure 4.

7) To find the x-intercept, solve $0 = \log_2 x + 3$ for x. $\log_2 x = -3 \implies$

$2^{-3} = x \implies x = \frac{1}{8}$; vertical asymptote : x = 0; See Figure 7.

10) Squaring x has the effect of reflecting the graph about the y-axis
since the domain is now all reals except 0. x-intercepts : ± 1,
vertical asymptote : x = 0; See Figure 10.

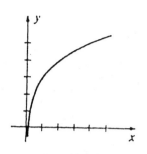

Figure 7

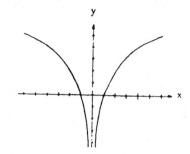

Figure 10

13) This is the same as the graph of $f(x) = \frac{1}{2}\log_2 x$. x-intercept : 1; vertical asymptote : $x = 0$; See Figure 13.

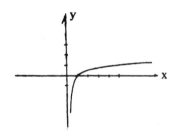

Figure 13

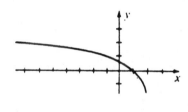

Figure 16

16) Think of the argument $2 - x$ as $-(x - 2)$. This has the effect of shifting the graph of $y = \log_3 x$ two units right and then reflecting that graph about the line $x = 2$. x-intercept : $0 = \log_3 (2 - x) \Rightarrow 2 - x = 3^0 \Rightarrow 2 - x = 1 \Rightarrow x = 1$; y-intercept : $f(0) = \log_3 2 \simeq 0.63$; vertical asymptote : $x = 2$; See Figure 16.

Exercises 4.4

19) The absolute value changes the
domain to all reals except 5.
x-intercepts : 4 and 6;
y-intercept : $\log_2 5 \simeq 2.32$;
vertical asymptote : $x = 5$;
See Figure 19.

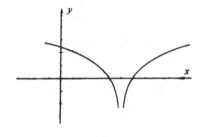

Figure 19

Exercises 4.5

1) $\log x = 3.6274 \Rightarrow x = 10^{3.6274} \simeq 4240.333$ or 4240 to three
significant figures

4) $\log x = 4.9680 \Rightarrow x = 10^{4.9680} \simeq 92{,}896.638$ or 92,900 to three
significant figures

7) $\ln x = 2.3 \Rightarrow x = e^{2.3} \simeq 9.974$ or 9.97 to three significant figures

10) $\ln x = 0.95 \Rightarrow x = e^{0.95} \simeq 2.59$ to three significant figures

13) (a) $I = 100\,I_0 \Rightarrow R = \log\left(\dfrac{100\,I_0}{I_0}\right) = \log 10^2 = 2$

(b) $I = 10{,}000\,I_0 \Rightarrow R = \log\left(\dfrac{10{,}000\,I_0}{I_0}\right) = \log 10^4 = 4$

(c) $I = 100{,}000\,I_0 \Rightarrow R = \log\left(\dfrac{100{,}000\,I_0}{I_0}\right) = \log 10^5 = 5$

15) $R = 2.3 \log (A + 34{,}000) - 7.5 \Rightarrow \dfrac{R + 7.5}{2.3} = \log (A + 34{,}000)$

$\Rightarrow A + 34{,}000 = 10^{\left(\frac{R + 7.5}{2.3}\right)} \Rightarrow A = 10^{\left(\frac{R + 7.5}{2.3}\right)} - 34{,}000$

16) See Exercise 15 above for the development of the formula for A (A_2
in Exercise 16). A_1 follows a similar pattern and the following
quotient is the desired formula. (continued on next page)

$$\frac{A_1}{A_2} = \frac{10^{\left(\frac{R+5.1}{2.3}\right)} - 3000}{10^{\left(\frac{R+7.5}{2.3}\right)} - 34000}$$

19) (a) vinegar : pH $\simeq -\log(6.3 \times 10^{-3}) = -(\log 6.3 + \log 10^{-3})$

   $= -(\log 6.3 - 3) = 3 - \log 6.3 \simeq 2.2$

   (b) carrots : pH $\simeq -\log(1.0 \times 10^{-5}) = -(\log 1.0 + \log 10^{-5})$

   $= -(\log 1.0 - 5) = 5 - \log 1.0 = 5$

   (c) sea water : pH $\simeq -\log(5.0 \times 10^{-9}) = -(\log 5.0 + \log 10^{-9})$

   $= -(\log 5.0 - 9) = 9 - \log 5.0 \simeq 8.3$

22) $1 < \text{pH} < 14 \Rightarrow 1 < -\log[H^+] < 14 \Rightarrow -1 > \log[H^+] > -14 \Rightarrow$

   $10^{-1} > 10^{\log[H^+]} > 10^{-14} \Rightarrow 10^{-14} < [H^+] < 10^{-1}$

25) $p = 29e^{-0.000034h} \Rightarrow \frac{p}{29} = e^{-0.000034h} \Rightarrow \ln\left(\frac{p}{29}\right) =$

   $-0.000034h \Rightarrow h \simeq -29{,}412 \ln\left(\frac{p}{29}\right) = 29{,}412 \ln\left(\frac{29}{p}\right)$

28) Substituting $v = 0$ and $m = m_1 + m_2$ in $v = -a \ln m + b$ yields

   $0 = -a \ln(m_1 + m_2) + b$ so $b = a \ln(m_1 + m_2)$. At burnout,

   $m = m_1$, so $v = -a \ln m_1 + b = -a \ln m_1 + a \ln(m_1 + m_2) =$

   $a[\ln(m_1 + m_2) - \ln m_1] = a \ln\left(\frac{m_1 + m_2}{m_1}\right).$

31) Let $q(t) = \frac{1}{2}q_0$. Now $\frac{1}{2}q_0 = q_0 e^{-0.0063t} \Rightarrow \frac{1}{2} = e^{-0.0063t}$

   $\Rightarrow -\ln 2 = -0.0063t \Rightarrow t = \frac{\ln 2}{0.0063} \simeq 110$ days

Exercises 4.5

34) $25{,}000 = 6000\,e^{0.1t}$ ➡ $\dfrac{25}{6} = e^{0.1t}$ ➡ $0.1t = \ln\left(\dfrac{25}{6}\right)$ ➡

$t = 10\ln\left(\dfrac{25}{6}\right) \simeq 14.27$ years or about 171 months

Exercises 4.6

1) (a) $10^x = 7$ ➡ $\log 10^x = \log 7$ ➡ $x = \log 7$ (b) 0.85

4) (a) $10^x = 6$ ➡ $\log 10^x = \log 6$ ➡ $x = \log 6$ (b) 0.78

7) (a) $3^{x+4} = 2^{1-3x}$ ➡ $\log 3^{x+4} = \log 2^{1-3x}$ ➡ $(x+4)\log 3 = $

$(1-3x)\log 2$ ➡ $x\log 3 + 4\log 3 = \log 2 - 3x\log 2$ ➡

$x\log 3 + 3x\log 2 = \log 2 - 4\log 3$ ➡ $x = \dfrac{\log 2 - 4\log 3}{\log 3 + 3\log 2}$ or

$\dfrac{\log\left(\frac{2}{81}\right)}{\log 24}$ (b) -1.16

10) (a) $2^{-x^2} = 5$ ➡ $\log 2^{-x^2} = \log 5$ ➡ $(-x^2)\log 2 = \log 5$ ➡

$-x^2 = \dfrac{\log 5}{\log 2}$ ➡ $x^2 = -\dfrac{\log 5}{\log 2}$; There are no real solutions for x

since $x^2 \geq 0$ and $-\dfrac{\log 5}{\log 2} < 0$.

13) (a) $\log(x^2 + 4) - \log(x + 2) = 3 + \log(x - 2)$ ➡

$\log\left(\dfrac{x^2 + 4}{x + 2}\right) - \log(x - 2) = 3$ ➡ $\log\left(\dfrac{x^2 + 4}{x^2 - 4}\right) = 3$ ➡

$\dfrac{x^2 + 4}{x^2 - 4} = 1000$ ➡ $4004 = 999x^2$ ➡ $x = \pm\sqrt{\dfrac{4004}{999}} = $

$\pm\dfrac{2}{3}\sqrt{\dfrac{1001}{111}}$; the negative solution is extraneous so

$x = \dfrac{2}{3}\sqrt{\dfrac{1001}{111}}$ is the only solution;   (b) 2.00

16) $\log \sqrt{x} = \sqrt{\log x} \implies \frac{1}{2} \log x = \sqrt{\log x} \implies \frac{1}{4} (\log x)^2 = \log x \implies$

$\frac{1}{4} (\log x)^2 - \log x = 0 \implies (\log x)(\frac{1}{4} \log x - 1) = 0 \implies \log x = 0, 4;$

$(\log x = 0 \implies x = \underline{1})$ and $(\log x = 4 \implies x = \underline{10,000})$

19) $x^{\sqrt{\log x}} = 10^8 \implies \log x^{\sqrt{\log x}} = \log 10^8 \implies \sqrt{\log x} (\log x) = 8$

$\implies (\log x)^{3/2} = 8 \implies \log x = 8^{2/3} = 4 \implies x = 10,000$

22) $y = \dfrac{10^x - 10^{-x}}{2} \implies 2y = 10^x - 10^{-x}$ {multiply by 10^x} $\implies$

$10^{2x} - 2y\, 10^x - 1 = 0$ {treat as a quadratic in 10^x} $\implies$

$10^x = \dfrac{2y \pm \sqrt{4y^2 + 4}}{2} = y \pm \sqrt{y^2 + 1};\ \sqrt{y^2 + 1} > y$ so

$y - \sqrt{y^2 + 1} < 0$, but $10^x > 0$ therefore $\log(y - \sqrt{y^2 + 1})$ is

extraneous and $x = \log(y + \sqrt{y^2 + 1})$

25) Use ln instead of log in the solution of Exercise 22.

28) $y = \dfrac{e^x + e^{-x}}{e^x - e^{-x}} \implies ye^x - ye^{-x} = e^x + e^{-x} \implies ye^{2x} - y = e^{2x} + 1$

$\implies (y - 1) e^{2x} = y + 1 \implies e^{2x} = \dfrac{y + 1}{y - 1} \implies e^x = \sqrt{\dfrac{y + 1}{y - 1}} \implies$

$x = \ln \left(\sqrt{\dfrac{y + 1}{y - 1}} \right) = \dfrac{1}{2} \ln \left(\dfrac{y + 1}{y - 1} \right)$

31) 50% of the light reaching a depth of 13 meters corresponds to the

equation $\frac{1}{2} I_0 = I_0 a^{13}$. Solving for a we have $a^{13} = \frac{1}{2}$ or $a = \sqrt[13]{\frac{1}{2}}$.

Now letting $I = 0.01 I_0$, $a = \sqrt[13]{\frac{1}{2}}$, and using the formula $x = \dfrac{\log \left(\frac{I}{I_0} \right)}{\log a}$

from the text, we have $x = \dfrac{\log(0.01)}{\log \sqrt[13]{\frac{1}{2}}} = \dfrac{-2}{\frac{1}{13}(\log 1 - \log 2)} =$

$\dfrac{-26}{(0 - \log 2)} = \dfrac{26}{\log 2} \simeq 86.4$ meters.

Exercises 4.6

34) The half-life of ^{131}I being 8 days corresponds to the equation $\frac{1}{2} A_0 = A_0 a^{-8}$. Solving for a we have $a^{-8} = \frac{1}{2} \Rightarrow a = (\frac{1}{2})^{-1/8} = 2^{1/8} = \sqrt[8]{2}. \{\approx 1.09\}$

Exercises 4.7

1) $\log_2 \frac{1}{16} = \log_2 2^{-4} = -4$

4) $10^{3 \log 2} = 10^{\log 2^3} = 2^3 = 8$

7) $\log_4 2 = \log_4 4^{1/2} = \frac{1}{2}$

10) $\log (\log 10^{10}) = \log (10) = 1$

13) This is the same graph as $y = (\frac{2}{3})^X$. y-intercept : 1; horizontal asymptote : $y = 0$; See Figure 13.

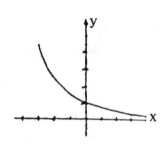

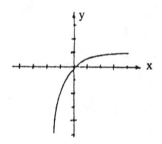

Figure 13   Figure 16

16) $f(x) = 1 - 3^{-X} = 1 - (\frac{1}{3})^X = - (\frac{1}{3})^X + 1$; This graph could be thought of taking $y = (\frac{1}{3})^X$, reflecting it about the x-axis { because of the "−" in front of $(\frac{1}{3})^X$ }, and then shifting that graph one unit up. y-intercept : 0; horizontal asymptote : $y = 1$; See Figure 16.

19) The 2 in front of $\log_3 x$ has the effect of doubling each y value of the graph of $y = \log_3 x$. x-intercept : 1; vertical asymptote : $x = 0$; See Figure 19.

22) y-intercept : $\frac{1}{2}$; horizontal asymptote : $y = 0$; See Figure 22.

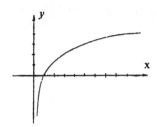

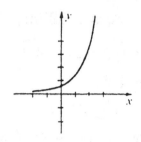

Figure 19   Figure 22

25) $\log_8 (x - 5) = \frac{2}{3} \Rightarrow x - 5 = 8^{2/3} = 4 \Rightarrow x = 9$

28) $\log \sqrt[4]{x + 1} = \frac{1}{2} \Rightarrow (x + 1)^{1/4} = 10^{1/2} \Rightarrow x + 1 = 10^2 \Rightarrow x = 99$

31) $2^{5x + 3} = 3^{2x + 1} \Rightarrow (5x + 3) \log 2 = (2x + 1) \log 3 \Rightarrow$

$5x \log 2 + 3 \log 2 = 2x \log 3 + \log 3 \Rightarrow 5x \log 2 - 2x \log 3 =$

$\log 3 - 3 \log 2 \Rightarrow x = \dfrac{\log 3 - 3 \log 2}{5 \log 2 - 2 \log 3}$ or $\dfrac{\log \left(\frac{3}{8}\right)}{\log \left(\frac{32}{9}\right)}$

34) $\ln x = 1 + \ln (x + 1) \Rightarrow \ln x - \ln (x + 1) = 1 \Rightarrow$

$\ln \left(\dfrac{x}{x + 1}\right) = 1 \Rightarrow \dfrac{x}{x + 1} = e^1 \Rightarrow x = e (x + 1) \Rightarrow x = ex + e \Rightarrow$

$x - ex = e \Rightarrow (1 - e) x = e \Rightarrow x = \dfrac{e}{1 - e}$

37) Use log instead of ln in the solution of problem 28, Exercises 4.6

40) (a) $\log x = -2.4260 \Rightarrow x = 10^{-2.4260} \simeq 0.00375$

    (b) $-2.4260 = -3 + 0.5740$; Using Table 1, 0.5740 is found in
       the row $N = 3.7$ under the column labeled 5; multiply 3.75 by
       10^{-3} to obtain 0.00375

43) (a) $x = \ln 6.6 \simeq 1.887$

(b) Using Table 3 with $n = 6$ under the column labeled 0.6, we have
$\ln 6.6 = 1.887$

46) $A = 1000 \left(1 + \dfrac{0.12}{4}\right)^{4(1)} \simeq \$1,125.51$

49) (a) $35000 = 10000e^{0.11t} \Rightarrow e^{0.11t} = 3.5 \Rightarrow 0.11t = \ln 3.5$

$\Rightarrow t = \dfrac{100 \ln 3.5}{11} \simeq 11.39$ years

(b) $20000 = 10000e^{0.11t} \Rightarrow e^{0.11t} = 2 \Rightarrow 0.11t = \ln 2$

$\Rightarrow t = \dfrac{100 \ln 2}{11} \simeq 6.30$ years

52) (a) $m = 6 - (2.5) \log \left(\dfrac{10^{0.4} L_0}{L_0}\right) = 6 - (2.5)(0.4) = 5$

(b) $m = 6 - (2.5) \log \left(\dfrac{L}{L_0}\right) \Rightarrow 2.5 \log \left(\dfrac{L}{L_0}\right) = 6 - m \Rightarrow$

$\log \left(\dfrac{L}{L_0}\right) = \dfrac{6 - m}{2.5} \Rightarrow \dfrac{L}{L_0} = 10^{\left(\frac{6 - m}{2.5}\right)} \Rightarrow L = L_0 \, 10^{\left(\frac{6 - m}{2.5}\right)}$

55) Let $y = \frac{1}{2} y_0$. $\frac{1}{2} y_0 = y_0 e^{-0.3821t} \Rightarrow \frac{1}{2} = e^{-0.3821t} \Rightarrow$

$\ln \frac{1}{2} = -0.3821t \Rightarrow t = \dfrac{-\ln 2}{-0.3821} \simeq 1.814$ years

There are other possible answers for Exercises 1, 4, 7, and 10. The answers listed are the smallest (in magnitude) two positive and negative coterminal angles.

1) $120^\circ + (1)\ 360^\circ = 480^\circ$;
   $120^\circ + (2)\ 360^\circ = 840^\circ$;
   $120^\circ - (1)\ 360^\circ = -240^\circ$;
   $120^\circ - (2)\ 360^\circ = -600^\circ$;
   See Figure 1.

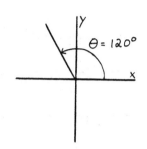

Figure 1

4) $315^\circ + (1)\ 360^\circ = 675^\circ$;
   $315^\circ + (2)\ 360^\circ = 1035^\circ$;
   $315^\circ - (1)\ 360^\circ = -45^\circ$;
   $315^\circ - (2)\ 360^\circ = -405^\circ$;
   See Figure 4.

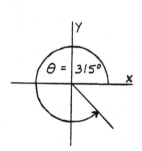

Figure 4

7) $620^\circ - (1)\ 360^\circ = 260^\circ$;
   $620^\circ + (1)\ 360^\circ = 980^\circ$;
   $620^\circ - (2)\ 360^\circ = -100^\circ$;
   $620^\circ - (3)\ 360^\circ = -460^\circ$;
   See Figure 7.

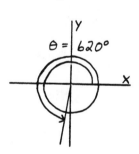

Figure 7

Exercises 5.1

10) $\frac{2\Pi}{3} + (1) \, 2\Pi = \frac{8\Pi}{3}$;

$\frac{2\Pi}{3} + (2) \, 2\Pi = \frac{14\Pi}{3}$;

$\frac{2\Pi}{3} - (1) \, 2\Pi = -\frac{4\Pi}{3}$;

$\frac{2\Pi}{3} - (2) \, 2\Pi = -\frac{10\Pi}{3}$;

See Figure 10.

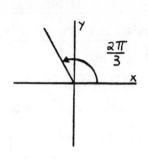

Figure 10

13) $150 \left(\frac{\Pi}{180}\right) = \frac{5\Pi}{6}$

16) $-135 \left(\frac{\Pi}{180}\right) = -\frac{3\Pi}{4}$

19) $450 \left(\frac{\Pi}{180}\right) = \frac{5\Pi}{2}$

22) $54 \left(\frac{\Pi}{180}\right) = \frac{3\Pi}{10}$

25) $\frac{2\Pi}{3} \left(\frac{180}{\Pi}\right) = 120^0$

28) $\frac{4\Pi}{3} \left(\frac{180}{\Pi}\right) = 240^0$

31) $-\frac{7\Pi}{2} \left(\frac{180}{\Pi}\right) = -630^0$

34) $9\Pi \left(\frac{180}{\Pi}\right) = 1620^0$

37) $2 \left(\frac{180}{\Pi}\right) \simeq 114.59156^0$; {subtract 114 and change to minutes}

$0.59156 \, (60) = 35.4936'$; {subtract 35 and change to seconds}

$0.4936 \, (60) \simeq 30''$; $\therefore$ 2 radians $\simeq 114^0 35' 30''$

{$\therefore$ is an abbreviation for "therefore"}

74

40) $4\left(\dfrac{180}{\Pi}\right) \simeq 229.18312^0$; {Solution is similar to that of Exercise

37} $0.18312 (60) = 10.9872'$; $0.9872 (60) \simeq 59''$;

$\therefore$ 4 radians $\simeq 229^0 10' 59''$

43) One minute is $\frac{1}{60}$ of one degree and one second is $\frac{1}{3600}$ of a degree.

Thus, $115^0 26' 27'' = (115 + \frac{26}{60} + \frac{27}{3600})^0 = (115 + \frac{1587}{3600})^0 \simeq 115.44^0$

46) Change the fractional portion of degrees to minutes. $0.864^0 =$

$(0.864) (60) = 51.84'$; Now change the fractional portion of

minutes to seconds. $0.84' = (0.84) (60) = 50.4''$;

$\therefore$ $12.864^0 \simeq 12^0 51' 50''$

49) (a) $\theta = \dfrac{s}{r} = \dfrac{7}{4} = 1.75$ radians;

(b) $\left(\dfrac{7}{4}\right)\left(\dfrac{180}{\Pi}\right) = \dfrac{315}{\Pi} \simeq 100.27^0$

52) $s = r\theta = (\frac{1}{2} \cdot 120)(2.2) = 132$ cm {Note that the radius is $\frac{1}{2} \cdot 120$,

<u>not</u> 120.}

55) radius $= \frac{1}{2} \cdot 8000 = 4000$ miles

(a) $s = r\theta = 4000\left(60 \cdot \dfrac{\Pi}{180}\right) = \dfrac{4000\Pi}{3} \simeq 4189$ miles

(b) $s = r\theta = 4000\left(45 \cdot \dfrac{\Pi}{180}\right) = 1000\Pi \simeq 3142$ miles

(c) $s = r\theta = 4000\left(30 \cdot \dfrac{\Pi}{180}\right) = \dfrac{2000\Pi}{3} \simeq 2094$ miles

(d) $s = r\theta = 4000\left(10 \cdot \dfrac{\Pi}{180}\right) = \dfrac{2000\Pi}{9} \simeq 698$ miles

(e) $s = r\theta = 4000\left(1 \cdot \dfrac{\Pi}{180}\right) = \dfrac{200\Pi}{9} \simeq 70$ miles

58) (a) $(2400)(2\Pi) = 4800\Pi$ radians per minute

Exercises 5.1

(b) $s = r\theta = (\frac{1}{2} \cdot 18)(4800\Pi) = 43,200\Pi$ inches per minute

$\qquad = 3600\Pi$ ft/min {divide 43,200 by 12}

61) The lengths of chain s_1 and s_2 are equal. $s_1 = r_1\theta_1$ and $s_2 = r_2\theta_2$

so $r_2\theta_2 = r_1\theta_1 \Rightarrow \theta_2 = \dfrac{r_1\theta_1}{r_2}$.

Exercises 5.2

Answers to Exercises 1, 4, 7, 10, 13, and 16 are in the order sin, cos, tan, csc, sec, cot.

1) $\frac{4}{5}, \frac{3}{5}, \frac{4}{3}, \frac{5}{4}, \frac{5}{3}, \frac{3}{4}$

4) $c = \sqrt{5^2 + 6^2} = \sqrt{61}$; $\dfrac{5\sqrt{61}}{61}, \dfrac{6\sqrt{61}}{61}, \dfrac{5}{6}, \dfrac{\sqrt{61}}{5}, \dfrac{\sqrt{61}}{6}, \dfrac{6}{5}$

7) $c = \sqrt{a^2 + b^2}$; $\dfrac{a\sqrt{a^2+b^2}}{a^2+b^2}, \dfrac{b\sqrt{a^2+b^2}}{a^2+b^2}, \dfrac{a}{b}, \dfrac{\sqrt{a^2+b^2}}{a}, \dfrac{\sqrt{a^2+b^2}}{b}, \dfrac{b}{a}$

10) $c = \sqrt{a^2 + a^2} = \sqrt{2a^2} = \sqrt{2}\,a$; $\dfrac{\sqrt{2}}{2}, \dfrac{\sqrt{2}}{2}, 1, \sqrt{2}, \sqrt{2}, 1$

13) opp. side $= \sqrt{6^2 - 5^2} = \sqrt{11}$; $\dfrac{\sqrt{11}}{6}, \dfrac{5}{6}, \dfrac{\sqrt{11}}{5}, \dfrac{6\sqrt{11}}{11}, \dfrac{6}{5}, \dfrac{5\sqrt{11}}{11}$

16) adj. side $= \sqrt{2^2 - (\sqrt{3})^2} = 1$; $\dfrac{\sqrt{3}}{2}, \dfrac{1}{2}, \sqrt{3}, \dfrac{2\sqrt{3}}{3}, 2, \dfrac{\sqrt{3}}{3}$

19) (a) 6.197 (b) 6.1970

22) both (a) and (b) are 0.3872

25) both (a) and (b) are 0.4718

28) both (a) and (b) are 0.9316

31) $\cos^2\theta + \sin^2\theta = 1 \Rightarrow \cos^2\theta = 1 - \sin^2\theta \Rightarrow$

$\cos\theta = \sqrt{1 - \sin^2\theta}$; Now $\sec\theta = \dfrac{1}{\cos\theta} = \dfrac{1}{\sqrt{1 - \sin^2\theta}}$;

{$\cos\theta = \pm\sqrt{1 - \sin^2\theta}$, the $\pm$ notation appears in a later section}

76

34) $\cot^2\theta + 1 = \csc^2\theta \Rightarrow \cot^2\theta = \csc^2\theta - 1 \Rightarrow \cot\theta = \sqrt{\csc^2\theta - 1}$

37) Refer to figure 5.14

(a) $\sin\theta = \dfrac{b}{c} = \dfrac{b/b}{c/b} = \dfrac{1}{\csc\theta}$

(b) $\cos\theta = \dfrac{a}{c} = \dfrac{a/a}{c/a} = \dfrac{1}{\sec\theta}$

(c) $\tan\theta = \dfrac{b}{a} = \dfrac{b/b}{a/b} = \dfrac{1}{\cot\theta}$

40) $\cos\theta = \sqrt{1 - \sin^2\theta} = \sqrt{1 - \dfrac{4}{25}} = \dfrac{\sqrt{21}}{5}$; $\quad \tan\theta = \dfrac{\sin\theta}{\cos\theta} = \dfrac{2/5}{\sqrt{21}/5}$

$= \dfrac{2\sqrt{21}}{21}$; $\quad \cot\theta = \dfrac{\sqrt{21}}{2}$; $\quad \sec\theta = \dfrac{5\sqrt{21}}{21}$; $\quad \csc\theta = \dfrac{5}{2}$

Answers to Exercises 43, 46, 49, 52, 55, and 58 are in the order
$\sin\theta$, $\cos\theta$, $\tan\theta$, $\cot\theta$, $\sec\theta$, and $\csc\theta$.

43) $x = 4$ and $y = -3 \Rightarrow r = \sqrt{4^2 + (-3)^2} = 5$;

$-\frac{3}{5}, \frac{4}{5}, -\frac{3}{4}, -\frac{4}{3}, \frac{5}{4}, -\frac{5}{3}$

46) $x = -1$ and $y = 2 \Rightarrow r = \sqrt{(-1)^2 + 2^2} = \sqrt{5}$;

$\dfrac{2\sqrt{5}}{5}, -\dfrac{\sqrt{5}}{5}, -2, -\dfrac{1}{2}, -\sqrt{5}, \dfrac{\sqrt{5}}{2}$

49) $2y - 7x + 2 = 0 \Rightarrow y = \frac{7}{2}x - 1$ so the slope of the given line is
$\frac{7}{2}$. The line through the origin with that slope is $y = \frac{7}{2}x$. If $x = -2$,
then $y = -7$ and $(-2, -7)$ is a point on the terminal side of θ.
$x = -2$ and $y = -7 \Rightarrow r = \sqrt{(-2)^2 + (-7)^2} = \sqrt{53}$;

$-\dfrac{7\sqrt{53}}{53}, -\dfrac{2\sqrt{53}}{53}, \dfrac{7}{2}, \dfrac{2}{7}, -\dfrac{\sqrt{53}}{2}, -\dfrac{\sqrt{53}}{7}$

52) The equation of the line bisecting the third quadrant is $y = x$. If
$x = -1$, then $y = -1$ and $(-1, -1)$ is a point on the terminal side of

Exercises 5.2

θ. $x = -1$ and $y = -1 \Rightarrow r = \sqrt{(-1)^2 + (-1)^2} = \sqrt{2}$.

$-\dfrac{\sqrt{2}}{2}, -\dfrac{\sqrt{2}}{2}, 1, 1, -\sqrt{2}, -\sqrt{2}$

55) For $\theta = 180°$, choose $x = -1$ and $y = 0$. r is 1.

   0, -1, 0, undefined, -1, undefined

58) For $\theta = \dfrac{5\Pi}{2}$, choose $x = 0$ and $y = 1$. r is 1.

   1, 0, undefined, 0, undefined, 1

61) $\cos \theta > 0$ is true in quadrants I and IV; $\sin \theta < 0$ is true in
   quadrants III and IV; ∴ both are true in quadrant IV

64) $\sec \theta > 0$ is true in quadrants I and IV; $\tan \theta < 0$ is true in
   quadrants II and IV; ∴ both are true in quadrant IV

67) $\sec \theta < 0$ is true in quadrants II and III; $\tan \theta > 0$ is true in
   quadrants I and III; ∴ both are true in quadrant III

70) $\cos \theta < 0$ is true in quadrants II and III; $\csc \theta < 0$ is true in
   quadrants III and IV; ∴ both are true in quadrant III

Exercises 5.3

In Exercises 1, 4, 7, and 10, answers for part (b) are in the order
sin t, cos t, tan t, csc t, sec t, and cot t.

1) 2Π is one revolution (counterclockwise)   (a) (1, 0)

   (b) 0, 1, 0, undefined, 1, undefined

4) $\dfrac{5\Pi}{2}$ is $1\frac{1}{4}$ revolutions (counterclockwise)   (a) (0, 1)

   (b) 1, 0, undefined, 1, undefined, 0

7) $-\dfrac{7\Pi}{2}$ is $1\frac{3}{4}$ revolutions (clockwise)   (a) (0, 1)

   (b) 1, 0, undefined, 1, undefined, 0

10) $-\dfrac{9\Pi}{4}$ is $1\frac{1}{8}$ revolutions (clockwise) { See Example 1 (b) }

    (a) $\left[\dfrac{\sqrt{2}}{2}, -\dfrac{\sqrt{2}}{2}\right]$ (b) $-\dfrac{\sqrt{2}}{2}, \dfrac{\sqrt{2}}{2}, -1, -\sqrt{2}, \sqrt{2}, -1$

13) Since both coordinates are positive, $P(t) = (\frac{3}{5}, \frac{4}{5})$ is in the first

    quadrant and $0 < t < \dfrac{\Pi}{2}$.

    (a) $\Pi < t + \Pi < \dfrac{3\Pi}{2} \Rightarrow P(t + \Pi) = (-\frac{3}{5}, -\frac{4}{5})$ and is in Q III;

    (b) $-\Pi < t - \Pi < -\dfrac{\Pi}{2} \Rightarrow P(t - \Pi) = (-\frac{3}{5}, -\frac{4}{5})$ and is in Q III;

    (c) $-\dfrac{\Pi}{2} < -t < 0 \Rightarrow P(-t) = (\frac{3}{5}, -\frac{4}{5})$ and is in Q IV;

    (d) $-\dfrac{3\Pi}{2} < -t - \Pi < -\Pi \Rightarrow P(-t - \Pi) = (-\frac{3}{5}, \frac{4}{5})$ and is in Q II

16) Since $x < 0$ and $y > 0$, $P(t)$ is in the second quadrant and $\dfrac{\Pi}{2} < t < \Pi$.

    (a) $\dfrac{3\Pi}{2} < t + \Pi < 2\Pi \Rightarrow P(t + \Pi) = (\frac{15}{17}, -\frac{8}{17})$ and is in Q IV;

    (b) $-\dfrac{\Pi}{2} < t - \Pi < 0 \Rightarrow P(t - \Pi) = (\frac{15}{17}, -\frac{8}{17})$ and is in Q IV;

    (c) $-\Pi < -t < -\dfrac{\Pi}{2} \Rightarrow P(-t) = (-\frac{15}{17}, -\frac{8}{17})$ and is in Q III;

    (d) $-2\Pi < -t - \Pi < -\dfrac{3\Pi}{2} \Rightarrow P(-t - \Pi) = (\frac{15}{17}, \frac{8}{17})$ and is in Q I

19) $\sin\left(-\dfrac{\Pi}{2}\right) = -\sin\left(\dfrac{\Pi}{2}\right) = -1$; $\cos\left(-\dfrac{\Pi}{2}\right) = \cos\left(\dfrac{\Pi}{2}\right) = 0$

22) Let $f(t) = \tan t$. Now $f(-t) = \tan(-t) = -\tan t$ and $-f(t) = -\tan t$, implying that f is odd. The graph of f is symmetric with respect to the origin.

25) We know the sine function has period 2Π, that is, $\sin(t + 2\Pi) = \sin t$ for all t. Now $\csc(t + 2\Pi) = \dfrac{1}{\sin(t + 2\Pi)} = \dfrac{1}{\sin t} = \csc t$ which implies the cosecant function has period of at most 2Π. Now

Exercises 5.3

assume csc $(t + k) = \csc t$ for all t where $0 < k < 2\Pi$. If $t = 0$, then csc $k = \csc 0$, which is undefined. The only value k (where $0 < k < 2\Pi$) for which csc k is undefined is $k = \Pi$. Thus, csc $(t + \Pi)$ would have to equal csc t for all t. Now let $t = \frac{\Pi}{2}$.

csc $\left(\frac{\Pi}{2} + \Pi \right) = -1$ and csc $\frac{\Pi}{2} = 1$, which is a contradiction. Therefore 2Π is the smallest positive real number for which the cosecant is periodic and is thus the period of the cosecant.

27) Let a be any real number. If $\tan t = \frac{y}{x} = a$ where $x^2 + y^2 = 1$, then

$a = \frac{\pm\sqrt{1 - x^2}}{x}$. Solving for x, we have $xa = \pm\sqrt{1 - x^2} \Rightarrow$

$x^2 a^2 = 1 - x^2 \Rightarrow x^2 (a^2 + 1) = 1 \Rightarrow x = \pm\frac{1}{\sqrt{a^2 + 1}}$. If $a > 0$,

choose the point P (x, y) on U where $x = \frac{1}{\sqrt{a^2 + 1}}$. If $a < 0$, choose

the point P (x, y) on U where $x = \frac{-1}{\sqrt{a^2 + 1}}$. If $a = 0$, choose P to be

$(1, 0)$. Thus, there is always a point P (x, y) on U such that $\frac{y}{x} = a$.

28) Let b be any nonzero real number and $b = \frac{1}{a}$ where a is the number

in Exercise 27. If $b = 0$, choose P to be $(0, 1)$. Thus, there is

always a point P (x, y) on U such that $\frac{x}{y} = b$.

Exercises 5.4

1) (a) Since t is in Q II, $t' = \Pi - t = \Pi - \frac{3\Pi}{4} = \frac{\Pi}{4}$;

(b) Since t is in Q III, $t' = t - \Pi = \frac{4\Pi}{3} - \Pi = \frac{\Pi}{3}$;

(c) For $-\dfrac{\Pi}{6}$, use the coterminal angle between 0 and 2Π, $\dfrac{11\Pi}{6}$;

Since $\dfrac{11\Pi}{6}$ is in Q IV, $t' = 2\Pi - t = 2\Pi - \dfrac{11\Pi}{6} = \dfrac{\Pi}{6}$

4) (a) For $\dfrac{8\Pi}{3}$, use the coterminal angle between 0 and 2Π, $\dfrac{2\Pi}{3}$;

Since $\dfrac{2\Pi}{3}$ is in Q II, $t' = \Pi - t = \Pi - \dfrac{2\Pi}{3} = \dfrac{\Pi}{3}$;

(b) Since $\dfrac{7\Pi}{4}$ is in Q IV, $t' = 2\Pi - t = 2\Pi - \dfrac{7\Pi}{4} = \dfrac{\Pi}{4}$;

(c) For $-\dfrac{7\Pi}{6}$, use the coterminal angle between 0 and 2Π, $\dfrac{5\Pi}{6}$;

Since $\dfrac{5\Pi}{6}$ is in Q II, $t' = \Pi - t = \Pi - \dfrac{5\Pi}{6} = \dfrac{\Pi}{6}$

7) (a) Since 240^0 is in Q III, $\theta' = \theta - 180^0 = 240^0 - 180^0 = 60^0$;

(b) Since 340^0 is in Q IV, $\theta' = 360^0 - \theta = 360^0 - 340^0 = 20^0$;

(c) For -110^0, use the coterminal angle between 0^0 and 360^0, 250^0;
Since 250^0 is in Q III, $\theta' = \theta - 180^0 = 250^0 - 180^0 = 70^0$

10) (a) Since $335^0\ 20'$ is in Q IV, $\theta' = 360^0 - \theta = 360^0 - 335^0\ 20' = 24^0\ 40'$;

(b) For -620^0, use the coterminal angle between 0^0 and 360^0, 100^0;
Since 100^0 is in Q II, $\theta' = 180^0 - \theta = 180^0 - 100^0 = 80^0$;

(c) For $-185^0\ 40'$, use the coterminal angle between 0^0 and 360^0, $174^0\ 20'$; Since $174^0\ 20'$ is in Q II, $\theta' = 180^0 - \theta = 180^0 - 174^0\ 20' = 5^0\ 40'$

13) (a) $\tan\left(-\dfrac{5\Pi}{4}\right) = -\tan\left(\dfrac{5\Pi}{4}\right) = -\tan\dfrac{\Pi}{4} = -1$;

(b) $\cot 315^0 = -\cot 45^0 = -1$

Exercises 5.4

16) (a) $\tan(-135°) = -\tan 135° = \tan 45° = 1$;

(b) $\sec 225° = -\sec 45° = -\sqrt{2}$

19) $\cot 189° 10' = \cot 9° 10' = 6.197$ from Table 4; Using a calculator, $\cot 189° 10' \simeq 6.197028$

22) $\tan 207° 10' = \tan 27° 10' = 0.5132$ from Table 4; Using a calculator, $\tan 207° 10' \simeq 0.51319497$

Calculator Exercises 5.4

1) Find INV COS of 0.8620 {30.45811}, subtract the integer portion {30} from that number {30.45811}, multiply the remaining number {0.45811} by 60 to find the number of minutes {27.49}, and round that result {27}; $\theta \simeq 30°27'$ or its fourth quadrant reference angle 329°33'

4) INV COS (0.8) = 36.869898; 60 (0.869898) $\simeq$ 52; $\theta \simeq 36°52'$ or its fourth quadrant reference angle 323°8'

7) INV TAN (-1.456) = -55.518207; 60 (-0.518207) $\simeq$ -31; $\theta \simeq -55° 31'$; Using its coterminal angle between 0° and 360°, we have $-55° 31' + 360° = 304° 29'$ or its second quadrant reference angle 124° 29'

10) Since the diameter is 2 km, the radius is 1 km and the circumference is 2Π km.

(a) $\frac{500}{2\Pi} \simeq 79.577472$ revolutions; $(2\Pi)(0.577472) \simeq$ 3.6283607 radians; Now $x = \cos(3.6283607) \simeq -0.8838$ and $y = \sin(3.6283607) \simeq -0.4678$.

(b) $x = \cos 2 \simeq -0.4161$ and $y = \sin 2 \simeq 0.9093$

1) (a) The secant function increases on the intervals

$$\left[-2\Pi, -\frac{3\Pi}{2}\right), \left(-\frac{3\Pi}{2}, -\Pi\right], \left[0, \frac{\Pi}{2}\right), \text{ and } \left(\frac{\Pi}{2}, \Pi\right].$$

(b) The secant function decreases on the intervals

$$\left[-\Pi, -\frac{\Pi}{2}\right), \left(-\frac{\Pi}{2}, 0\right], \left[\Pi, \frac{3\Pi}{2}\right), \text{ and } \left(\frac{3\Pi}{2}, 2\Pi\right].$$

(c) As x approaches $\Pi/2$ through values less than $\Pi/2$, the secant increases without bound ($\rightarrow \infty$). As x approaches $\Pi/2$ through values greater than $\Pi/2$, the secant decreases without bound ($\rightarrow -\infty$).

(d) Let f (x) = sec x. Then f (-x) = sec (-x) = sec x = f (x) so f is an even function and is symmetric with respect to the y-axis.

4) Examine Figures 5.35, 5.36, and 5.37 paying particular attention to minimum and maximum points, vertical asymptotes, and the relationship to the sine and cosine for the cosecant and the secant graphs. For the cotangent graph, remember the x-intercepts, the vertical asymptotes, and the fact that the cotangent is always decreasing over any interval on which it is defined.

7) Each y value of $f(x) = \frac{1}{2}\sin x$ is $\frac{1}{2}$ of those on the graph of $f(x) = \sin x$. x-intercepts : Πn, where n is any integer; See Figure 7.

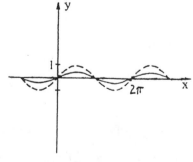

Figure 7

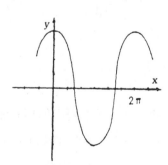

Figure 10

Exercises 5.5

10) Each y value of $f(x) = 4\cos x$ is 4 times those on the graph of $f(x) = \cos x$. x-intercepts : $\frac{\Pi}{2} + \Pi n$; See Figure 10.

13) Each y value of $f(x) = -\sin x$ is -1 times the y values of $f(x) = \sin x$. This has the effect of revolving the graph of $f(x) = \sin x$ about the x-axis. x-intercepts : Πn See Figure 13.

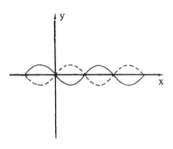

Figure 13

Calculator Exercises 5.5

1)

x	$\dfrac{\sin x}{x}$
0.1	0.9983
0.01	1.0000
0.001	1.0000
0.0001	1.0000

From the results in the table, we see that as x gets close to 0 through values greater than 0, $f(x) = \dfrac{\sin x}{x}$ gets very close to 1.

4)

x	$\dfrac{\sin x}{1 + \cos x}$
0.1	0.0500
0.01	0.0050
0.001	0.0005
0.0001	0.0001

From the results in the table, we see that as x gets close to 0 through values greater than 0, $f(x) = \dfrac{\sin x}{1 + \cos x}$ gets very close to 0.

Exercises 5.6

1) For (a) - (h), the amplitude is the multiplier in front of the sine function. The period is found by dividing 2Π by the multiplier of x.

(a) amplitude : 4; period : 2Π;   (b) amplitude : 1; period : Π/2;

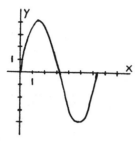

Figure 1 (a)

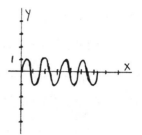

Figure 1 (b)

(c) amplitude : $\frac{1}{4}$; period : 2Π;   (d) amplitude : 1; period : 8Π;

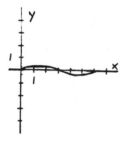

Figure 1 (c)

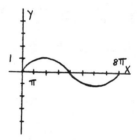

Figure 1 (d)

(e) amplitude : 2; period : 8Π;   (f) amplitude : $\frac{1}{2}$; period : Π/2;

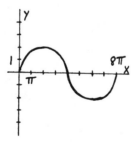

Figure 1 (e)

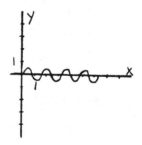

Figure 1 (f)

Exercises 5.6

(g) amplitude : 4; period : 2Π;   (h) amplitude : 1; period : Π/2;

$$f(x) = \sin(-4x) = -\sin 4x$$

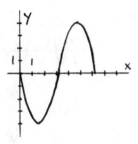

Figure 1 (g)   Figure 1 (h)

4) This problem follows a pattern similar to Exercise 1. Listed below are the amplitudes and periods for Exercise 3 and 4.

(a) 3, 2Π   (b) 1, 2Π/3   (c) $\frac{1}{3}$, 2Π   (d) 1, 6Π

(e) 2, 6Π   (f) $\frac{1}{3}$, Π   (g) 3, 2Π   (h) 1, 2Π/3

7) amplitude : 3; phase shift : Π/2; See Figure 7.

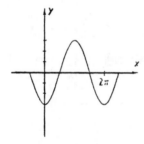

Figure 7   Figure 10

10) phase shift : Π/3; See Figure 10.

13) period : 2Π/3; phase shift : - Π/3; See Figure 13.

16) amplitude : 3; period : 2Π/3; phase shift : Π/3; See Figure 16.

19) amplitude : 6; period : 2; See Figure 19.

22) amplitude : 4; period : $\frac{2}{3}$; See Figure 22.

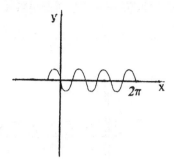

Figure 13

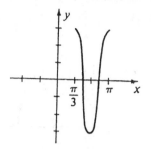

Figure 16

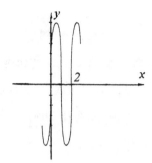

Figure 19

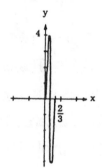

Figure 22

25) $y = \frac{1}{2} \sec x$ has y values that
are $\frac{1}{2}$ those of $y = \sec x$; The
vertical asymptote positions
remain the same as
$y = \sec x$; See Figure 25.

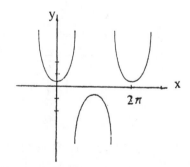

Figure 25

28) The -3 in front of the $\sec x$ has the effect of multiplying each y
value of $y = \sec x$ by 3 and reflecting that graph about the x-axis.
Vertical asymptotes do not change. See Figure 28.

31) phase shift : $\Pi/4$; This phase shift will move the vertical

asymptotes of $y = \csc x$ from their normal position, $x = \Pi n$, to $x = \Pi/4 + \Pi n$. See Figure 31.

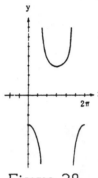

Figure 28

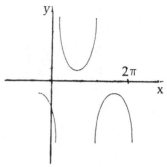

Figure 31

34) period : $\Pi/2$; The effect is to change the asymptotes from $x = ..., 0, \Pi, 2\Pi, ...$ to $x = ..., 0, \Pi/2, \Pi, ...$; To find the period for the tangent and cotangent functions, we divide Π by the multiplier of x, rather than 2Π. See Figure 34.

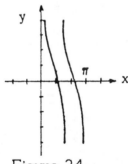

Figure 34

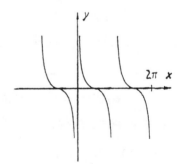

Figure 37

37) $y = \frac{1}{2} \cot x$ has y values that are $\frac{1}{2}$ those of $y = \cot x$. The effect is to vertically compress the graph of $y = \cot x$. See Figure 37.

40) This is the graph of $y = \cot x$ reflected about the x-axis. See Figure 40.

43) phase shift : $- \Pi/2$; See Figure 43.

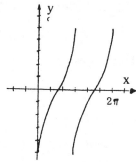

Figure 40

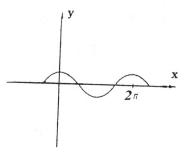

Figure 43

46) (a) amplitude = 2 since the y values range from −2 to 2; the horizontal distance between two minimum points is 4 so the period is 4;

(b) $\frac{2\Pi}{b} = 4 \Rightarrow b = \frac{\Pi}{2}$; $\frac{-c}{\frac{\Pi}{2}} = 1 \Rightarrow c = -\frac{\Pi}{2}$; $y = 2\sin\left(\frac{\Pi}{2}x - \frac{\Pi}{2}\right)$

(c) the phase shift of 1 unit was used to compute the value of c in in part (b)

49) There are two cycles per second so the period is $\frac{1}{2}$.

$\frac{2\Pi}{b} = \frac{1}{2} \Rightarrow b = 4\Pi$

52) (a) $\frac{2\Pi}{b} = 23 \Rightarrow b = \frac{2\Pi}{23}$; similarly, $b = \frac{2\Pi}{28}$ and $b = \frac{2\Pi}{33}$

(b) physical : $y = \sin\left(\frac{2\Pi}{23}x\right) = \sin\left(\frac{2\Pi}{23}\cdot 7660\right) \approx 0.27$ or 27%

emotional : $y = \sin\left(\frac{2\Pi}{28}x\right) \approx -0.43$ or −43%

intellectual : $y = \sin\left(\frac{2\Pi}{33}x\right) \approx 0.69$ or 69%

Exercises 5.7

For Exercises 1, 4, 7, and 10, it is often helpful to examine the zeros
of the functions when using addition of y-coordinates to sketch graphs.
i.e. When one function is zero, y will have the value of the other
function.

1) $y = \cos x$ when $x = \Pi n$ {the zeros of $3 \sin x$} and $y = 3 \sin x$ when
   $x = \Pi/2 + \Pi n$ {the zeros of $\cos x$}; y-intercept : 1; See Figure 1.

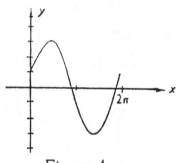

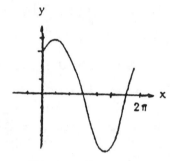

Figure 1   Figure 4

4) $y = 2 \sin x$ when $x = \Pi/2 + \Pi n$ and $y = 3 \cos x$ when $x = \Pi n$;
   y-intercept : 3; See Figure 4.

7) $y = \cos x$ when $x = \Pi n$ and $y = -\sin x$ when $x = \Pi/2 + \Pi n$;
   y-intercept : 1; See Figure 7.

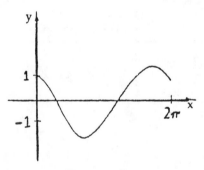

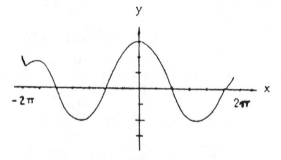

Figure 7   Figure 10

10) $y = 2 \cos x$ when $x = (2n + 1)\Pi$ {since the period of $\cos \frac{1}{2}x$ is 4Π}
    and $y = \cos \frac{1}{2}x$ when $x = \Pi/2 + \Pi n$; y-intercept : 3 See Figure 10.

13) y-intercept : 0; The graph will look like the graph of $y = \frac{1}{2}x$.
Adding the value of $\sin x$ { which ranges from -1 to 1} to the value
of $\frac{1}{2}x$ will have little effect on y as x gets large { positively or
negatively}. See Figure 13.

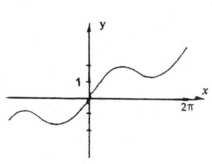

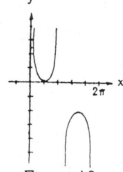

Figure 13   Figure 16

16) The -1 has the effect of lowering the graph of $y = \csc x$ by 1 unit.
The range is $(-\infty, -2] \cup [0, \infty)$. See Figure 16.

19) The -1 has the effect of lowering the graph of $y = \cos x$ by 1 unit.
The range is $[-2, 0]$. See Figure 19.

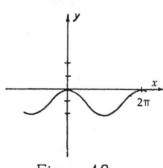

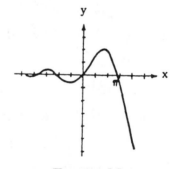

Figure 19   Figure 22

22) There will be zeros every Πn units since the sine is zero there.
Every time the sine is 1, the graph will touch the graph of $y = 2^x$.
Every time the sine is -1, the graph will touch the graph of $y =$
-2^x. Refer to Example 4 for additional discussion. See Figure 22.

25) It is helpful to graph $y = \frac{1}{2} x^2$ and $y = -\frac{1}{2} x^2$ on the same coordinate axes for this problem. Plot x-intercepts at every Πn units. Plot points on the graph of $y = \frac{1}{2} x^2$ every $\dfrac{(4n + 1)\Pi}{2}$ units {this is where $\sin x = 1$}. Plot points on the graph of $y = -\frac{1}{2} x^2$ at every $\dfrac{(4n - 1)\Pi}{2}$ units {this is where $\sin x = -1$}. The curve will now oscillate between the plotted points. See Figure 25.

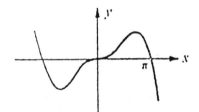

Figure 25

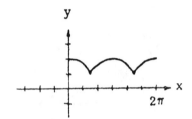

Figure 28

28) Graph $y = \cos x$. Now graph $y = |\cos x|$, the absolute value will "flip" all negative portions of the graph of $y = \cos x$ about the x-axis. Finally, add 1 to all y values. This shifts $y = |\cos x|$ up one unit. See Figure 28.

31) The "+12" added to D (t) means the sine curve will be centered on $y = 12$ instead of $y = 0$ {the x-axis}. Now K/2 determines the variation from 12. In this case, the graph will vary from 6 to 18, whereas the example in the text {Figure 5.53} varied from 9 to 15. See Figure 31.

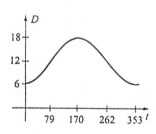

Figure 31

92

34) (a) A high of 20°C and a low of 10°C imply that d = 15 and a = 5. The period is 24 so $b = \frac{\Pi}{12}$. Assuming that the high temperature occurs at 6 P.M. and not at 6 A.M., the average temperature of 15°C (before the high of 20°C) would occur at noon which corresponds to t = 12. Letting this correspond to the first zero of the sine function, we have $f(t) = 5 \sin\left[\frac{\Pi}{12}(t - 12)\right] + 15$.

(b) See Figure 34 (b).

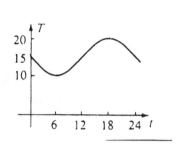

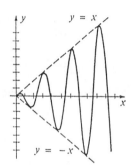

Figure 34 (b)   Figure 37

37) f (x) = x cos 2x has zeros every (Π/4) + (Π/2)n units. It will touch y = x every Πn units since cos 2x = 1 ⟹ 2x = 2Πn ⟹ x = Πn. It will touch y = -x every (Π/2) + Πn units since cos 2x = -1 ⟹ 2x = Π + 2Πn ⟹ x = (Π/2) + Πn. See Figure 37.

1) $\beta = 90° - \alpha = 60°$; $c = 20 \sec 30° = \frac{40\sqrt{3}}{3} \simeq 23$;

$a = \sqrt{c^2 - b^2} = \frac{20\sqrt{3}}{3} \simeq 12$

4) $\beta = 90° - \alpha = 53°$; $c = b \sec \alpha = 24 \sec 37° \simeq 30$;

$a = c \sin \alpha \simeq 18$

Exercises 5.8

7) $\alpha = 90^0 - \beta = 18^09'$; $c = b \sec \alpha = 240.0 \sec 18^09' \simeq 252.6$;

   $a = c \sin \alpha \simeq 78.7$

10) $c = \sqrt{a^2 + b^2} = \sqrt{1042} \simeq 32$; $\tan \alpha = \frac{31}{9} \Rightarrow \alpha \simeq 74^0$;

   $\beta = 90^0 - \alpha \simeq 16^0$

13) $\beta = 90^0 - \alpha = 52^014'$; $c = b \sec \alpha = 512.0 \sec 37^046' \simeq 647.7$;

   $a = c \sin \alpha \simeq 396.7$

16) $a = \sqrt{c^2 - b^2} = \sqrt{932.2} \simeq 30.5$; $\cos \alpha = \frac{21.6}{37.4} \Rightarrow \alpha \simeq 54^043'$;

   $\beta = 90^0 - \alpha \simeq 35^017'$

19) $\sin 54^020' = \frac{x}{85} \Rightarrow x \simeq 69.1$; height $= x + 1.5 \simeq 70.6$ m.

22) Let x be the distance from the bottom of the ladder to the building
and let h be the height of the top of the ladder on the building.

   $\sin 22^0 = \frac{x}{20} \Rightarrow x \simeq 7.5$ ft.; $\cos 22^0 = \frac{h}{20} \Rightarrow h \simeq 18.5$ ft.;

   Now $x = 20 \sin 22^0 + 3$ and $h = \sqrt{20^2 - x^2} \simeq 17.0$ ft. so the top of
the ladder moved 1.5 ft down { 18.5 to 17.0}. See Figure 22.

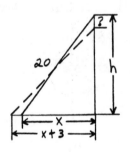

Figure 22

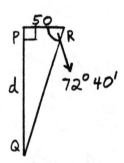

Figure 25

25) $\tan 72^040' = \frac{d}{50} \Rightarrow d \simeq 160$ m; See Figure 25.

28) $\sin \alpha = \frac{5}{24} \Rightarrow \alpha \simeq 12^0$; See Figure 28.

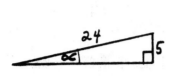

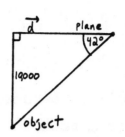

Figure 28   Figure 31

31) Let d be the distance traveled. $\tan 42^0 = \dfrac{10000}{d} \implies$

d = 10000 cot 42⁰; converting this to mph, we have

$\dfrac{10000 \cot 42^0 \text{ ft}}{1 \text{ minute}} \cdot \dfrac{60 \text{ minutes}}{1 \text{ hour}} \cdot \dfrac{1 \text{ mile}}{5280 \text{ ft}} \approx 126.2$ mph;

See Figure 31.

34) $\tan \theta = \dfrac{h}{d} \implies h = d \tan \theta$; h = 1000 tan 59⁰ ≈ 1664 m.

See Figure 34.

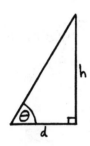

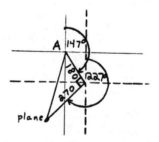

Figure 34   Figure 37

37) The plane's flight forms a right triangle with legs 180 {½ · 360}

miles and 270 {¾ · 360} miles. The distance from A to the

airplane is then $\sqrt{180^2 + 270^2} \approx 324.5$ mi. See Figure 37.

95

Exercises 5.8

40) Let x_1 be the horizontal length of the lower section and x_2 be the horizontal length of the upper section. $\tan 25^0 = \dfrac{15}{x_1} \Rightarrow x_1 = $ 15 cot 25^0; $\tan 35^0 = \dfrac{15}{x_2} \Rightarrow x_2 = 15$ cot 35^0; Let s_1 be the slide length of the lower section, s_2 be the slide length of the upper section, and s_3 be the slide length of the middle section.

$\sin 25^0 = \dfrac{15}{s_1} \Rightarrow s_1 = 15$ csc 25^0; $\sin 35^0 = \dfrac{15}{s_2} \Rightarrow$

$s_2 = 15$ csc 35^0. The middle section has a slide length of $s_3 = 100 - x_1 - x_2$. The total slide length is $s_1 + s_2 + s_3 \simeq 35.5 + 26.2 + 46.4 = 108.1$ ft.

43) The diagonal of the base is $\sqrt{8^2 + 6^2} = 10$. Tan $\theta = \frac{4}{10} \Rightarrow$
$\theta \simeq 21.8^0$

46) Form a right triangle with P, the center of the sun (C), and the point at the top of the sun (R). $\angle PCR = \frac{1}{2} (32') = 16'$; Let r be the radius of the sun. Tan $(\angle PCR) = \tan (0^0 \, 16') = \dfrac{r}{92,900,000} \Rightarrow$
$r \simeq 432,379$ mi. Thus, the diameter is $2r$ or $\simeq 864,758$ mi.

Calculator Exercises 5.8

1) $\beta = 48.73^0$; $c = a$ csc $\alpha \simeq 476.9$; $\qquad b = c \cos \alpha \simeq 358.5$
4) $\beta = 72.31^0$; $c = b$ sec $\alpha \simeq 1.372$; $\qquad a = c \sin \alpha \simeq 0.4169$

7) $c = \sqrt{a^2 + b^2} \simeq 48.67$; $\tan \alpha = \frac{a}{b} \Rightarrow \alpha \simeq 74.36^0$; $\beta \simeq 15.64^0$

10) $b = \sqrt{c^2 - a^2} \simeq 32.90$; $\sin \alpha = \frac{a}{c} \Rightarrow \alpha \simeq 69.53^0$; $\beta \simeq 20.47^0$

1) (a) $\omega = \left(\dfrac{2\Pi \text{ radians}}{\text{revolution}}\right) \cdot \left(\dfrac{100 \text{ revolutions}}{\text{minute}}\right) = 200\Pi \text{ radians/min}$

(b) The diameter of the wheel is 40 cm so its radius is 20 cm.

$x = 20 \cos (200\Pi t) \text{ cm}, \quad y = 20 \sin (200\Pi t) \text{ cm}$

4) Amplitude, $\frac{1}{3}$cm.; period $= \dfrac{2\Pi}{\Pi/4} = 8$ sec; frequency $= \dfrac{\Pi/4}{2\Pi} = \dfrac{1}{8}$

oscillation/second; The point is at $d = \frac{1}{3}$ cm when $t = 0$. It then

decreases in height until $(\Pi/4)\,t = \Pi$ or $t = 4$ where it obtains a

minimum of $d = -\frac{1}{3}$. It then reverses direction and increases to a

height of $d = \frac{1}{3}$ when $(\Pi/4)\,t = 2\Pi$ or $t = 8$ to complete one

oscillation.

7) period $= 3 \Rightarrow \dfrac{2\Pi}{\omega} = 3 \Rightarrow \omega = \dfrac{2\Pi}{3}$; amplitude $= 5 \Rightarrow a = 5$;

$d = 5 \cos\left(\dfrac{2\Pi}{3}t\right)$

10) The phase shift of the current

(I) is $\dfrac{-\Pi/3}{120\,\Pi} = -\dfrac{1}{360}$. The

current leads the emf (E) by

$\frac{1}{360}$ second. See Figure 10.

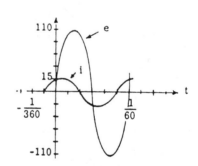

Figure 10

13) A high of 12 feet and a low of 3 feet imply that we have an average

height of 7.5 feet with an amplitude of 4.5. We have minimums at

8 A.M. and 8 P.M. so the period is 12 hours. $12 = \dfrac{2\Pi}{b} \Rightarrow b = \dfrac{\Pi}{6}$.

The average before a high occurs at $t = 10$ and we want this to

(continued on next page)

Exercises 5.9

correspond to the first zero of the sine curve. Combining all of the above, we have $y = (4.5) \sin \left[\frac{\Pi}{6} (t - 10) \right] + 7.5$.

Exercises 5.10

1) $330 \left(\frac{\Pi}{180} \right) = \frac{11\Pi}{6}$; $\quad 405 \left(\frac{\Pi}{180} \right) = \frac{9\Pi}{4}$; $\quad -150 \left(\frac{\Pi}{180} \right) = -\frac{5\Pi}{6}$;

$240 \left(\frac{\Pi}{180} \right) = \frac{4\Pi}{3}$; $\quad 36 \left(\frac{\Pi}{180} \right) = \frac{\Pi}{5}$

4) $s = r\theta = \left(15 \cdot \frac{1}{2} \right) \left(70 \cdot \frac{\Pi}{180} \right) = \frac{35\Pi}{12} \approx 9.16$ cm

7) (a) II   (b) III   (c) IV

10) $t' = \frac{5\Pi}{4} - \Pi = \frac{\Pi}{4}$; Use the coterminal angle $\frac{7\Pi}{6}$ for $-\frac{5\Pi}{6}$;

$t' = \frac{7\Pi}{6} - \Pi = \frac{\Pi}{6}$; Use the coterminal angle $\frac{7\Pi}{8}$ for $-\frac{9\Pi}{8}$;

$t' = \Pi - \frac{7\Pi}{8} = \frac{\Pi}{8}$

13) Answers are in the order sin, cos, tan, csc, sec, and cot.

(a) $x = 30$ and $y = -40 \Rightarrow r = \sqrt{30^2 + (-40)^2} = 50$;

$-\frac{4}{5}, \frac{3}{5}, -\frac{4}{3}, -\frac{5}{4}, \frac{5}{3}, -\frac{3}{4}$

(b) $2x + 3y + 6 = 0 \Rightarrow y = -\frac{2}{3}x - 2$ so the slope of the given line is $-\frac{2}{3}$. The line through the origin with that slope is $y = -\frac{2}{3}x$. If $x = -3$, then $y = 2$ and $(-3, 2)$ is a point on the terminal side of θ. $x = -3$ and $y = 2 \Rightarrow r = \sqrt{(-3)^2 + 2^2} = \sqrt{13}$;

$\frac{2\sqrt{13}}{13}, -\frac{3\sqrt{13}}{13}, -\frac{2}{3}, \frac{\sqrt{13}}{2}, -\frac{\sqrt{13}}{3}, -\frac{3}{2}$

(c) For $\theta = -90^0$, choose $x = 0$ and $y = -1$. r is 1.

$-1, 0,$ undefined, $-1,$ undefined, 0

16) amplitude : $\frac{2}{3}$; period : 2Π; See Figure 16.

19) amplitude : 3; period : $\Pi/2$; See Figure 19.

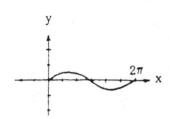

Figure 16

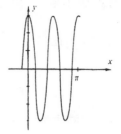

Figure 19

22) period : Π; See Figure 22.

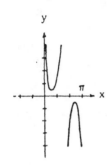

Figure 22

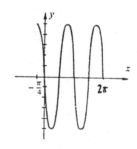

Figure 25

25) amplitude : 5; period : Π; phase shift : – Π/4; See Figure 25.

28) The cosine makes the graph
of $y = 1 + x + \cos x$ oscillate
about the line $y = 1 + x$;
y-intercept : 2;
See Figure 28.

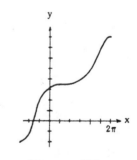

Figure 28

31) $\beta = 90^0 - 54^0 40' = 35^0 20'$; $c = b \sec \alpha = 220 \sec 54^0 40' \simeq 380$;
$a = c \sin \alpha \simeq 310$

Exercises 5.10

34) (a) $545 \text{ rpm} = \dfrac{545 \cdot 2\Pi}{60} \text{ radians/sec} \simeq 57 \text{ rad/sec}$

(b) $d = 22.625 \text{ feet} \implies c = \Pi d = 22.625\Pi \text{ feet.}$

$$\left(\frac{22.625\Pi \text{ ft}}{\text{rev.}}\right)\left(\frac{545 \text{ rev.}}{\text{min.}}\right)\left(\frac{1 \text{ mile}}{5280 \text{ ft}}\right)\left(\frac{60 \text{ min.}}{1 \text{ hour}}\right) \simeq 440.2 \text{ mph}$$

37) The angle ϕ has an adjacent side of $\frac{1}{2}$ (230 m) or 115 m.

$\tan \phi = \frac{147}{115} \implies \phi \simeq 52^0$

40) (a) Let h be the height of the building and x the distance between the

two buildings. $\tan 59^0 = \dfrac{h - 50}{x}$ and $\tan 62^0 = \dfrac{h}{x} \implies x =$

$(h - 50) \cot 59^0$ and $x = h \cot 62^0 \implies h \cot 59^0 - 50 \cot 59^0 =$

$h \cot 62^0 \implies h = \dfrac{50 \cot 59^0}{\cot 59^0 - \cot 62^0} \simeq 434.5 \text{ ft.}$

(b) From part (a), $x = h \cot 62^0 \simeq 231.0 \text{ ft.}$

1) $\cos\theta\,\sec\theta = \cos\theta\left(\dfrac{1}{\cos\theta}\right) = 1$

4) $\sin\alpha\,\cot\alpha = \sin\alpha\left(\dfrac{\cos\alpha}{\sin\alpha}\right) = \cos\alpha$

7) $(1 + \cos\alpha)(1 - \cos\alpha) = 1 - \cos^2\alpha = \sin^2\alpha$

10) $(\tan\theta + \cot\theta)\tan\theta = \tan^2\theta + 1 = \sec^2\theta$

13) $(1 + \sin\alpha)(1 - \sin\alpha) = 1 - \sin^2\alpha = \cos^2\alpha = \dfrac{1}{\sec^2\alpha}$

16) $\dfrac{\sin w + \cos w}{\cos w} = \dfrac{\sin w}{\cos w} + \dfrac{\cos w}{\cos w} = \tan w + 1$

19) $\sin t\,(\csc t - \sin t) = 1 - \sin^2 t = \cos^2 t$

22) $\cos\theta\,(\tan\theta + \cot\theta) = \cos\theta \cdot \dfrac{\sin\theta}{\cos\theta} + \cos\theta \cdot \dfrac{\cos\theta}{\sin\theta} = \sin\theta + \dfrac{\cos^2\theta}{\sin\theta}$

$= \dfrac{\sin^2\theta + \cos^2\theta}{\sin\theta} = \dfrac{1}{\sin\theta} = \csc\theta$

25) $(\cos^2 x - 1)(\tan^2 x + 1) = \cos^2 x\,\tan^2 x - \tan^2 x + \cos^2 x - 1 =$

$\sin^2 x - \tan^2 x + \cos^2 x - 1 = -\tan^2 x = -(\sec^2 x - 1) = 1 - \sec^2 x$

28) $2\csc^2 y - 1 = \dfrac{2}{\sin^2 y} - 1 = \dfrac{2 - \sin^2 y}{\sin^2 y} = \dfrac{1 + \cos^2 y + \sin^2 y - \sin^2 y}{\sin^2 y}$

$= \dfrac{1 + \cos^2 y}{\sin^2 y}$

31) $\dfrac{1 + \cos t}{\sin t} + \dfrac{\sin t}{1 + \cos t} = \dfrac{(1 + \cos t)^2 + \sin^2 t}{\sin t\,(1 + \cos t)} =$

$\dfrac{1 + 2\cos t + \cos^2 t + \sin^2 t}{\sin t\,(1 + \cos t)} = \dfrac{2 + 2\cos t}{\sin t\,(1 + \cos t)} = \dfrac{2\,(1 + \cos t)}{\sin t\,(1 + \cos t)} =$

$\dfrac{2}{\sin t} = 2\csc t$

34) $\dfrac{\sec\theta + \csc\theta}{\sec\theta - \csc\theta} = \dfrac{\dfrac{1}{\cos\theta} + \dfrac{1}{\sin\theta}}{\dfrac{1}{\cos\theta} - \dfrac{1}{\sin\theta}} = \dfrac{\dfrac{\sin\theta + \cos\theta}{\cos\theta\,\sin\theta}}{\dfrac{\sin\theta - \cos\theta}{\cos\theta\,\sin\theta}} = \dfrac{\sin\theta + \cos\theta}{\sin\theta - \cos\theta}$

Exercises 6.1

37) $\dfrac{1 + \csc \beta}{\sec \beta} - \cot \beta = \dfrac{1 + \csc \beta - \cot \beta \sec \beta}{\sec \beta} = \dfrac{1 + \csc \beta - \csc \beta}{\sec \beta} =$

$\dfrac{1}{\sec \beta} = \cos \beta$

40) $\dfrac{\cot \theta - \tan \theta}{\sin \theta + \cos \theta} = \dfrac{\dfrac{\cos \theta}{\sin \theta} - \dfrac{\sin \theta}{\cos \theta}}{\sin \theta + \cos \theta} = \dfrac{\dfrac{\cos^2 \theta - \sin^2 \theta}{\cos \theta \sin \theta}}{\sin \theta + \cos \theta} =$

$\dfrac{(\cos \theta + \sin \theta)(\cos \theta - \sin \theta)}{\cos \theta \sin \theta (\sin \theta + \cos \theta)} = \dfrac{\cos \theta}{\cos \theta \sin \theta} - \dfrac{\sin \theta}{\cos \theta \sin \theta} =$

$\csc \theta - \sec \theta$

43) $\csc^4 t - \cot^4 t = (\csc^2 t + \cot^2 t)(\csc^2 t - \cot^2 t) = \csc^2 t + \cot^2 t$

$\{1 + \cot^2 t = \csc^2 t \Rightarrow \csc^2 t - \cot^2 t = 1\}$

46) $\dfrac{1}{\csc y - \cot y} \cdot \dfrac{\csc y + \cot y}{\csc y + \cot y} = \dfrac{\csc y + \cot y}{\csc^2 y - \cot^2 y} = \csc y + \cot y$

49) $\dfrac{\cot u - 1}{\cot u + 1} = \dfrac{(\cot u - 1)/\cot u}{(\cot u + 1)/\cot u} = \dfrac{1 - \tan u}{1 + \tan u}$

52) $\sin^4 \theta + 2 \sin^2 \theta \cos^2 \theta + \cos^4 \theta = (\sin^2 \theta + \cos^2 \theta)^2 = 1^2 = 1$

55) $(\sec t + \tan t)^2 = \left(\dfrac{1}{\cos t} + \dfrac{\sin t}{\cos t}\right)^2 = \left(\dfrac{1 + \sin t}{\cos t}\right)^2 = \dfrac{(1 + \sin t)^2}{\cos^2 t} =$

$\dfrac{(1 + \sin t)^2}{1 - \sin^2 t} = \dfrac{(1 + \sin t)^2}{(1 + \sin t)(1 - \sin t)} = \dfrac{1 + \sin t}{1 - \sin t}$

58) $\dfrac{\sin t}{1 - \cos t} \cdot \dfrac{1 + \cos t}{1 + \cos t} = \dfrac{\sin t (1 + \cos t)}{1 - \cos^2 t} = \dfrac{\sin t (1 + \cos t)}{\sin^2 t} =$

$\dfrac{1 + \cos t}{\sin t} = \csc t + \cot t$

61) $\left(\dfrac{\sin^2 x}{\tan^4 x}\right)^3 \left(\dfrac{\csc^3 x}{\cot^6 x}\right)^2 = \left(\dfrac{\sin^6 x}{\tan^{12} x}\right) \left(\dfrac{\csc^6 x}{\cot^{12} x}\right) = \dfrac{(\sin x \csc x)^6}{(\tan x \cot x)^{12}} = \dfrac{1^6}{1^{12}} = 1$

64) $(\csc t - \cot t)^4 (\csc t + \cot t)^4 = [(\csc t - \cot t)(\csc t + \cot t)]^4 =$

$(\csc^2 t - \cot^2 t)^4 = (1)^4 = 1$

67) $\dfrac{(\sin \alpha \cos \beta + \cos \alpha \sin \beta)/\cos \alpha \cos \beta}{(\cos \alpha \cos \beta - \sin \alpha \sin \beta)/\cos \alpha \cos \beta} = \dfrac{\tan \alpha + \tan \beta}{1 - \tan \alpha \tan \beta}$

70) $\sqrt{\dfrac{(1 - \sin \theta)(1 + \sin \theta)}{(1 + \sin \theta)(1 + \sin \theta)}} = \sqrt{\dfrac{1 - \sin^2 \theta}{(1 + \sin \theta)^2}} = \sqrt{\dfrac{\cos^2 \theta}{(1 + \sin \theta)^2}} =$

$\left| \dfrac{\cos \theta}{1 + \sin \theta} \right| = \dfrac{|\cos \theta|}{1 + \sin \theta}$ since $0 \le (1 + \sin \theta) \le 2$

73) $\dfrac{1}{\tan \beta + \cot \beta} = \dfrac{1}{\dfrac{\sin \beta}{\cos \beta} + \dfrac{\cos \beta}{\sin \beta}} = \dfrac{1}{\dfrac{\sin^2 \beta + \cos^2 \beta}{\cos \beta \sin \beta}} = \sin \beta \cos \beta$

76) $\sin^3 t + \cos^3 t = (\sin t + \cos t)(\sin^2 t - \sin t \cos t + \cos^2 t) =$

$(\sin t + \cos t)(1 - \sin t \cos t)$

79) $\dfrac{\tan x}{1 - \cot x} + \dfrac{\cot x}{1 - \tan x} = \dfrac{\dfrac{\sin x}{\cos x}}{1 - \dfrac{\cos x}{\sin x}} + \dfrac{\dfrac{\cos x}{\sin x}}{1 - \dfrac{\sin x}{\cos x}} =$

$\dfrac{\sin^2 x}{\cos x (\sin x - \cos x)} + \dfrac{\cos^2 x}{\sin x (\cos x - \sin x)} = \dfrac{\sin^3 x - \cos^3 x}{\cos x \sin x (\sin x - \cos x)}$

$= \dfrac{(\sin x - \cos x)(\sin^2 x + \sin x \cos x + \cos^2 x)}{\cos x \sin x (\sin x - \cos x)} =$

$\dfrac{1 + \sin x \cos x}{\cos x \sin x} = 1 + \sec x \csc x$

82) $\dfrac{\cot (-v)}{\csc (-v)} = \dfrac{-\cot v}{-\csc v} = \cos v$

85) $\ln (\cot x) = \ln (\tan x)^{-1} = -\ln (\tan x)$

88) $\ln |\csc x - \cot x| = \ln \left| \dfrac{(\csc x - \cot x)(\csc x + \cot x)}{\csc x + \cot x} \right| =$

$\ln \left| \dfrac{\csc^2 x - \cot^2 x}{\csc x + \cot x} \right| = \ln \left| \dfrac{1}{\csc x + \cot x} \right| = \ln |1| - \ln |\csc x + \cot x|$

$= -\ln |\csc x + \cot x|$

91) Choose any t so that $\sin t < 0$. Try $t = \dfrac{3\Pi}{2}$. $(1 \ne -1)$

Exercises 6.1

94) Try $t = \frac{\Pi}{6}$. $\left(\log 2 \neq \frac{1}{-\log 2} \right)$

97) Choose any t so that $\sec t \neq 2\Pi n$. Try $t = \frac{\Pi}{4}$. $(\cos \sqrt{2} \neq 1)$

100) $3 \cos^2 \theta + \cos \theta - 2 = 0 \Rightarrow (3 \cos \theta - 2)(\cos \theta + 1) = 0$

   Choose any θ so that $\cos \theta \neq \frac{2}{3}$, -1. Try $\theta = 0$. $(2 \neq 0)$

Note : In Exercises 101-112, assume $a > 0$ so that $\sqrt{a^2} = |a| = a$.

For Exercises 101-104, use $\sqrt{a^2 - x^2} = a \cos \theta$ since $\sqrt{a^2 - x^2} =$
$\sqrt{a^2 - a^2 \sin^2 \theta} = \sqrt{a^2 (1 - \sin^2 \theta)} = \sqrt{a^2 \cos^2 \theta} = |a||\cos \theta| = a \cos \theta$
since $\cos \theta > 0$ if $-\frac{\Pi}{2} < \theta < \frac{\Pi}{2}$.

103) $\dfrac{x^2}{\sqrt{a^2 - x^2}} = \dfrac{a^2 \sin^2 \theta}{a \cos \theta} = a \left(\dfrac{\sin \theta}{\cos \theta} \right) \sin \theta = a \tan \theta \sin \theta$

For Exercises 105-108, use $\sqrt{a^2 + x^2} = a \sec \theta$ since $\sqrt{a^2 + x^2} =$
$\sqrt{a^2 + a^2 \tan^2 \theta} = \sqrt{a^2 (1 + \tan^2 \theta)} = \sqrt{a^2 \sec^2 \theta} = |a||\sec \theta| = a \sec \theta$
since $\sec \theta > 0$ if $-\frac{\Pi}{2} < \theta < \frac{\Pi}{2}$.

106) $\dfrac{1}{\sqrt{a^2 + x^2}} = \dfrac{1}{a \sec \theta} = \dfrac{1}{a} \cos \theta$

For Exercises 109-112, use $\sqrt{x^2 - a^2} = a \tan \theta$ since $\sqrt{x^2 - a^2} =$
$\sqrt{a^2 \sec^2 \theta - a^2} = \sqrt{a^2 (\sec^2 \theta - 1)} = \sqrt{a^2 \tan^2 \theta} = |a||\tan \theta| = a \tan \theta$
since $\tan \theta > 0$ if $0 < \theta < \frac{\Pi}{2}$.

109) $\sqrt{x^2 - a^2} = a \tan \theta$

112) $\dfrac{\sqrt{x^2 - a^2}}{x^2} = \dfrac{a \tan \theta}{a^2 \sec^2 \theta} = \dfrac{1}{a} \left(\dfrac{\sin \theta}{\cos \theta} \right) \cos^2 \theta = \dfrac{1}{a} \sin \theta \cos \theta$

For Exercises 1, 4, 7, 10, 13, and 16, n denotes any integer.

1) $2\cos t + 1 = 0 \Rightarrow \cos t = -\frac{1}{2} \Rightarrow t = \frac{2\Pi}{3} + 2\Pi n, \frac{4\Pi}{3} + 2\Pi n$

4) $4\cos\theta - 2 = 0 \Rightarrow \cos\theta = \frac{1}{2} \Rightarrow \theta = \frac{\Pi}{3} + 2\Pi n, \frac{5\Pi}{3} + 2\Pi n$

7) $\sec^2\alpha - 4 = 0 \Rightarrow \sec^2\alpha = 4 \Rightarrow \sec\alpha = \pm 2 \Rightarrow \alpha = \frac{\Pi}{3} + \Pi n,$
$\frac{2\Pi}{3} + \Pi n$

10) $4\sin^2 x - 3 = 0 \Rightarrow \sin^2 x = \frac{3}{4} \Rightarrow \sin x = \pm\frac{\sqrt{3}}{2} \Rightarrow x = \frac{\Pi}{3} + \Pi n,$
$\frac{2\Pi}{3} + \Pi n$

13) $(2\sin\theta + 1)(2\cos\theta + 3) = 0 \Rightarrow \sin\theta = -\frac{1}{2} \text{ or } \cos\theta = -\frac{3}{2} \Rightarrow$
$\theta = \frac{7\Pi}{6} + 2\Pi n, \frac{11\Pi}{6} + 2\Pi n \quad \{\cos\theta = -\frac{3}{2} \text{ has no solutions}\}$

16) $\tan\alpha + \tan^2\alpha = 0 \Rightarrow \tan\alpha(\tan\alpha + 1) = 0 \Rightarrow \tan\alpha = 0, -1 \Rightarrow$
$\alpha = \Pi n \text{ or } \alpha = \frac{3\Pi}{4} + \Pi n$

19) $2\sin^2 u = 1 - \sin u \Rightarrow 2\sin^2 u + \sin u - 1 = 0 \Rightarrow$
$(2\sin u - 1)(\sin u + 1) = 0 \Rightarrow \sin u = \frac{1}{2}, -1 \Rightarrow$
$u = \frac{\Pi}{6}, \frac{5\Pi}{6}, \frac{3\Pi}{2}; \quad 30°, 150°, 270°$

22) $\sec\beta\csc\beta = 2\csc\beta \Rightarrow \sec\beta\csc\beta - 2\csc\beta = 0 \Rightarrow$
$\csc\beta(\sec\beta - 2) = 0 \Rightarrow \csc\beta = 0 \text{ or } \sec\beta = 2 \Rightarrow$
$\beta = \frac{\Pi}{3}, \frac{5\Pi}{3}; \quad 60°, 300° \quad \{\csc\beta = 0 \text{ has no solutions}\}$

25) $\sin^2\theta + \sin\theta - 6 = 0 \Rightarrow (\sin\theta + 3)(\sin\theta - 2) = 0 \Rightarrow$
$\sin\theta = -3, 2; \quad$ There are <u>no solutions</u> for either equation.

28) $\cos\theta - \sin\theta = 1 \implies \cos\theta = 1 + \sin\theta \implies \cos^2\theta = 1 + 2\sin\theta + \sin^2\theta \implies 1 - \sin^2\theta = 1 + 2\sin\theta + \sin^2\theta \implies 2\sin^2\theta + 2\sin\theta = 0 \implies 2\sin\theta(\sin\theta + 1) = 0 \implies \sin\theta = 0, -1 \implies$

$\theta = 0, \Pi, \dfrac{3\Pi}{2}$; $0°, 180°, 270°$; Since each side of the equation was squared, the solutions must be checked in the original equation. Checking 0, we have $1 = 1$. Checking Π, we have $-1 = 1$. Checking $\dfrac{3\Pi}{2}$, we have $1 = 1$. We conclude that Π is an extraneous solution.

31) $\tan\theta + \sec\theta = 1 \implies \sec^2\theta = (1 - \tan\theta)^2 \implies 1 + \tan^2\theta = 1 - 2\tan\theta + \tan^2\theta \implies 2\tan\theta = 0 \implies \theta = 0, \Pi$; $0°, 180°$; Checking 0, we have $1 = 1$. Checking Π, we have $-1 = 1$. We conclude that Π is an extraneous solution.

34) $\sec^5\theta = 4\sec\theta \implies \sec\theta(\sec^4\theta - 4) = 0 \implies \sec\theta = 0$ or $\sec^2\theta = \pm 2 \implies \sec\theta = \pm\sqrt{2}$ {since $\sec\theta \neq 0$ and $\sec^2\theta \neq -2$}

$\implies \theta = \dfrac{\Pi}{4}, \dfrac{3\Pi}{4}, \dfrac{5\Pi}{4}, \dfrac{7\Pi}{4}$; $45°, 135°, 225°, 315°$

37) $\sin^2 t - 4\sin t + 1 = 0 \implies \sin t = \dfrac{4 \pm \sqrt{16 - 4}}{2} = 2 \pm \sqrt{3}$;

$2 + \sqrt{3}$ is not in the range of the sine $\{-1 \text{ to } 1\}$, so $\sin t = 2 - \sqrt{3}$ $\{\simeq 0.2679\}$ which implies that $t = 15° 30'$ or $164° 30'$ {to the nearest 10 minutes}

40) $5\cos^2\alpha + 3\cos\alpha - 2 = 0 \implies (5\cos\alpha - 2)(\cos\alpha + 1) = 0 \implies \cos\alpha = \tfrac{2}{5}, -1 \implies \alpha = 66°30', 293°30', 180°$

43) $I = \dfrac{I_M}{2} \implies \dfrac{I_M}{2} = I_M \sin^3\left(\dfrac{\Pi t}{D}\right) \implies \sin^3\left(\dfrac{\Pi t}{D}\right) = \dfrac{1}{2} \implies$

$$\sin\left(\frac{\Pi t}{D}\right) = \sqrt[3]{\frac{1}{2}} \implies \frac{\Pi t}{D} \simeq 0.9169 \text{ and } 2.2247 \implies t \simeq 0.29 D$$

and $0.71 D$ {2.2247 is the reference number for 0.9169 in QII}

46) (a) On the surface, $x = 0$. Then $T = T_0 \sin(\omega t)$. Since the period is

24 hours, $24 = \frac{2\Pi}{\omega}$ or $\omega = \frac{\Pi}{12}$. The formula for the temperature

at the surface is then $T = T_0 \sin\left(\frac{\Pi}{12}t\right)$. T will be a minimum

when $\sin\left(\frac{\Pi}{12}t\right)$ equals -1. $\sin\left(\frac{\Pi}{12}t\right) = -1 \implies$

$\frac{\Pi}{12}t = \frac{3\Pi}{2} + 2\Pi n \implies t = 18 + 24n$ for $n = 0, 1, 2, \ldots$

(b) If $\lambda = 2.5$ and $x = 1$, then $T = T_0 e^{-2.5} \sin\left(\frac{\Pi}{12}t - 2.5\right)$. As in

part (a), $\sin\left(\frac{\Pi}{12}t - 2.5\right) = -1 \implies \frac{\Pi}{12}t - \frac{5}{2} = -\frac{\Pi}{2} + 2\Pi n \implies$

$\frac{\Pi}{12}t = \frac{5 - \Pi}{2} + 2\Pi n \implies t = \frac{6(5 - \Pi)}{\Pi} + 24n$ for $n = 0, 1,$

2, ... { If $\frac{3\Pi}{2}$ is used instead of $-\frac{\Pi}{2}$, then $n = -1, 0, 1, 2, \ldots$}

(c) This problem will be stated as follows in the next edition of the
text. As currently stated, it cannot be solved. "The maximum
temperature at the surface is T_0. At what depth is the maximum
temperature one-half of the surface maximum temperature?"
One-half of the surface maximum temperature is $\frac{1}{2}T_0$. The
maximum temperature will occur when $\sin(\omega t - \lambda x) = 1$, so now

$T = T_0 e^{-\lambda x}$. With $\lambda = 2.5$, we have $T_0 e^{-2.5x} = \frac{1}{2}T_0$. Solving

for x, $\frac{1}{2} = e^{-2.5x} \implies -\ln 2 = -2.5x \implies x = \frac{\ln 2}{2.5} \simeq 0.277$ feet

Exercises 6.3

1) (a) $\sin 46°37' = \cos (90° - 46°37') = \cos 43°23'$

(b) $\cos 73°12' = \sin (90° - 73°12') = \sin 16°48'$

(c) $\tan \dfrac{\Pi}{6} = \cot \left(\dfrac{\Pi}{2} - \dfrac{\Pi}{6}\right) = \cot \dfrac{\Pi}{3}$

(d) $\sec 17.28° = \csc (90° - 17.28°) = \csc 72.72°$

4) (a) $\sin \dfrac{\Pi}{12} = \cos \left(\dfrac{\Pi}{2} - \dfrac{\Pi}{12}\right) = \cos \dfrac{5\Pi}{12}$

(b) $\cos (0.64) = \sin \left(\dfrac{\Pi}{2} - 0.64\right) \quad \{ \simeq \sin 0.93\}$

(c) $\tan \sqrt{2} = \cot \left(\dfrac{\Pi}{2} - \sqrt{2}\right) \quad \{ \simeq \cot 0.16\}$

(d) $\sec (1.2) = \csc \left(\dfrac{\Pi}{2} - 1.2\right) \quad \{ \simeq \csc 0.37\}$

7) (a) $\tan 60° + \tan 225° = \sqrt{3} + 1$

(b) $\tan 285° = \tan (60° + 225°) = \dfrac{\tan 60° + \tan 225°}{1 - \tan 60° \tan 225°} =$

$\dfrac{\sqrt{3} + 1}{1 - (\sqrt{3})(1)} \cdot \dfrac{1 + \sqrt{3}}{1 + \sqrt{3}} = \dfrac{4 + 2\sqrt{3}}{-2} = -2 - \sqrt{3}$

10) (a) $\tan \dfrac{3\Pi}{4} - \tan \dfrac{\Pi}{6} = -1 - \dfrac{\sqrt{3}}{3} = \dfrac{-3 - \sqrt{3}}{3}$

(b) $\tan \dfrac{7\Pi}{12} = \tan \left(\dfrac{3\Pi}{4} - \dfrac{\Pi}{6}\right) = \dfrac{\tan \dfrac{3\Pi}{4} - \tan \dfrac{\Pi}{6}}{1 + \tan \dfrac{3\Pi}{4} \tan \dfrac{\Pi}{6}} = \dfrac{-1 - \dfrac{\sqrt{3}}{3}}{1 + (-1) \cdot \dfrac{\sqrt{3}}{3}} =$

$\dfrac{-3 - \sqrt{3}}{3 - \sqrt{3}} \cdot \dfrac{3 + \sqrt{3}}{3 + \sqrt{3}} = \dfrac{-12 - 6\sqrt{3}}{6} = -2 - \sqrt{3}$

13) $\cos 10° \sin 5° - \sin 10° \cos 5° = \sin (5° - 10°) = \sin (-5°)$

16) $\sin (-5) \cos 2 + \cos 5 \sin (-2) = \sin (-5) \cos (-2) + \cos (-5) \sin (-2)$

$= \sin (-2 + (-5)) = \sin (-7) \quad \{ \cos 2 = \cos (-2)$ since the cosine function is even$\}$

19) $\sin(\alpha + \beta) = \sin\alpha\cos\beta + \cos\alpha\sin\beta = \frac{7}{25} \cdot \frac{-3}{5} + \frac{-24}{25} \cdot \frac{-4}{5} = \frac{3}{5}$;

$\cos(\alpha + \beta) = \cos\alpha\cos\beta - \sin\alpha\sin\beta = \frac{-24}{25} \cdot \frac{-3}{5} - \frac{7}{25} \cdot \frac{-4}{5} = \frac{4}{5}$;

$\tan(\alpha + \beta) = \dfrac{\sin(\alpha + \beta)}{\cos(\alpha + \beta)} = \frac{3}{5} \div \frac{4}{5} = \frac{3}{4}$;

$\sin(\alpha - \beta) = \sin\alpha\cos\beta - \cos\alpha\sin\beta = \frac{7}{25} \cdot \frac{-3}{5} - \frac{-24}{25} \cdot \frac{-4}{5} = \frac{-117}{125}$;

$\cos(\alpha - \beta) = \cos\alpha\cos\beta + \sin\alpha\sin\beta = \frac{-24}{25} \cdot \frac{-3}{5} + \frac{7}{25} \cdot \frac{-4}{5} = \frac{44}{125}$;

$\tan(\alpha - \beta) = \dfrac{\sin(\alpha - \beta)}{\cos(\alpha - \beta)} = \frac{-117}{125} \div \frac{44}{125} = -\frac{117}{44}$

22) $\cos\left(x + \dfrac{\Pi}{2}\right) = \cos x \cos\dfrac{\Pi}{2} - \sin x \sin\dfrac{\Pi}{2} = -\sin x$

25) $\sin\left(\theta + \dfrac{\Pi}{4}\right) = \sin\theta\cos\dfrac{\Pi}{4} + \cos\theta\sin\dfrac{\Pi}{4} = \dfrac{\sqrt{2}}{2}\sin\theta + \dfrac{\sqrt{2}}{2}\cos\theta =$

$\dfrac{\sqrt{2}}{2}(\sin\theta + \cos\theta)$

28) $\tan\left(x - \dfrac{\Pi}{4}\right) = \dfrac{\tan x - \tan\dfrac{\Pi}{4}}{1 + \tan x \tan\dfrac{\Pi}{4}} = \dfrac{\tan x - 1}{1 + \tan x}$

31) $\sin(u + v) \cdot \sin(u - v) = (\sin u \cos v + \cos u \sin v) \cdot$

$(\sin u \cos v - \cos u \sin v) = \sin^2 u \cos^2 v - \cos^2 u \sin^2 v =$

$\sin^2 u (1 - \sin^2 v) - (1 - \sin^2 u) \sin^2 v =$

$\sin^2 u - \sin^2 u \sin^2 v - \sin^2 v + \sin^2 u \sin^2 v = \sin^2 u - \sin^2 v$

34) $\sin(u + v) + \sin(u - v) = \sin u \cos v + \cos u \sin v + \sin u \cos v -$

$\cos u \sin v = 2\sin u \cos v$

37) $\dfrac{\sin(u + v)}{\cos(u - v)} = \dfrac{(\sin u \cos v + \cos u \sin v)/\cos u \cos v}{(\cos u \cos v + \sin u \sin v)/\cos u \cos v} = \dfrac{\tan u + \tan v}{1 + \tan u \tan v}$

{We divide by (cos u cos v) to obtain the 1 in the denominator, the rest of the terms "fall in place"}

40) $\tan u + \tan v = \dfrac{\sin u}{\cos u} + \dfrac{\sin v}{\cos v} = \dfrac{\sin u \cos v + \cos u \sin v}{\cos u \cos v} = \dfrac{\sin(u + v)}{\cos u \cos v}$

Exercises 6.3

43) $\cot(u + v) = \dfrac{1}{\tan(u + v)} = \dfrac{1}{\dfrac{\tan u + \tan v}{1 - \tan u \tan v}} =$

$\dfrac{(1 - \tan u \tan v)/\tan u \tan v}{(\tan u + \tan v)/\tan u \tan v} = \dfrac{\cot u \cot v - 1}{\cot v + \cot u}$

46) (a) $\sec\left(\dfrac{\Pi}{2} - u\right) = \dfrac{1}{\cos\left(\dfrac{\Pi}{2} - u\right)} = \dfrac{1}{\sin u} = \csc u;$

(b) $\csc\left(\dfrac{\Pi}{2} - u\right) = \dfrac{1}{\sin\left(\dfrac{\Pi}{2} - u\right)} = \dfrac{1}{\cos u} = \sec u;$

(c) $\cot\left(\dfrac{\Pi}{2} - u\right) = \dfrac{1}{\tan\left(\dfrac{\Pi}{2} - u\right)} = \dfrac{1}{\cot u} = \tan u$

49) $\sin 4t \cos t = \sin t \cos 4t \implies \sin 4t \cos t - \sin t \cos 4t = 0 \implies$

$\sin(4t - t) = 0 \implies \sin 3t = 0 \implies 3t = \Pi n \implies t = \dfrac{\Pi}{3}n \;\{\text{or } 60°n\}$

for $n = 0, 1, \ldots, 5;\; \left\{0, \dfrac{\Pi}{3}, \dfrac{2\Pi}{3}, \Pi, \dfrac{4\Pi}{3}, \dfrac{5\Pi}{3}\right\}$

$\{0°, 60°, 120°, 180°, 240°, 360°\}$

52) $A = \sqrt{1^2 + (\sqrt{3})^2} = 2;\; \tan C = \dfrac{\sqrt{3}}{1} \implies C = \dfrac{\Pi}{3};$

$f(x) = 2\cos\left(4x - \dfrac{\Pi}{3}\right);\;$ amplitude $= 2$, period $= \dfrac{2\Pi}{4} = \dfrac{\Pi}{2},$

phase shift $= -\dfrac{-\Pi/3}{4} = \dfrac{\Pi}{12}$

55) (a) $y = 2\cos t + 3\sin t;\; A = \sqrt{2^2 + 3^2} = \sqrt{13};$

$\tan C = \tfrac{3}{2} \implies C \approx 0.98;\; y = \sqrt{13}\cos(t - C);$

amplitude $= \sqrt{13}$, period $= \dfrac{2\Pi}{1} = 2\Pi$

(b) $y = 0 \implies \cos(t - C) = 0 \implies t - C = \dfrac{\Pi}{2} + \Pi n \implies$

$t = C + \dfrac{\Pi}{2} + \Pi n \approx 2.5536 + \Pi n$ for $n = 0, 1, 2, \ldots$

58) $\sin\left[120\Pi t - \dfrac{\Pi}{2}\right] = \sin(120\Pi t)\cos\dfrac{\Pi}{2} - \cos(120\Pi t)\sin\dfrac{\Pi}{2} =$

$-\cos(120\Pi t)$; Now $y = -10\cos(120\Pi t) + 5\sin(120\Pi t)$;

$A = \sqrt{(-10)^2 + 5^2} = 5\sqrt{5}$; $\tan C = \dfrac{5}{-10} \implies C \approx -0.4636$;

$y \approx 5\sqrt{5}\cos(120\Pi t + 0.4636)$

1) $\sin\theta = \sqrt{1 - \cos^2\theta} = \dfrac{4}{5}$; $\sin 2\theta = 2\sin\theta\cos\theta = \dfrac{24}{25}$; $\cos 2\theta =$

$\cos^2\theta - \sin^2\theta = \dfrac{-7}{25}$; $\tan 2\theta = \dfrac{\sin 2\theta}{\cos 2\theta} = -\dfrac{24}{7}$

4) $\cos\theta = \sqrt{1 - \sin^2\theta} = \dfrac{3}{5}$; $\sin 2\theta = 2\sin\theta\cos\theta = -\dfrac{24}{25}$; $\cos 2\theta =$

$\cos^2\theta - \sin^2\theta = \dfrac{-7}{25}$; $\tan 2\theta = \dfrac{\sin 2\theta}{\cos 2\theta} = \dfrac{24}{7}$

7) $\tan\theta = 1$ and $-180° < \theta < -90° \implies \theta = -135°$; $\cos\theta = -\dfrac{\sqrt{2}}{2}$;

$-90° < \dfrac{\theta}{2} < -45°$ indicates that $\dfrac{\theta}{2}$ is in Q IV; $\sin\dfrac{\theta}{2} = -\sqrt{\dfrac{1 - \cos\theta}{2}}$

$= -\sqrt{(1 + \sqrt{2}/2)/2} = -\sqrt{(2 + \sqrt{2})/4} = -\sqrt{2 + \sqrt{2}}/2$;

$\cos\dfrac{\theta}{2} = \sqrt{\dfrac{1 + \cos\theta}{2}} = \sqrt{(1 - \sqrt{2}/2)/2} = \sqrt{(2 - \sqrt{2})/4} =$

$\sqrt{2 - \sqrt{2}}/2$; $\tan\dfrac{\theta}{2} = \dfrac{1 - \cos\theta}{\sin\theta} = \dfrac{\dfrac{2 + \sqrt{2}}{2}}{-\sqrt{2}/2} = -\dfrac{2 + \sqrt{2}}{\sqrt{2}} =$

$-(\sqrt{2} + 1)$

Exercises 6.4

10) (a) $\cos 165^0 = -\sqrt{\dfrac{1 + \cos 330^0}{2}} = -\sqrt{(1 + \sqrt{3}/2)/2} =$

$-\sqrt{(2 + \sqrt{3})/4} = -\sqrt{2 + \sqrt{3}}/2;$

(b) $\sin 157^0 30' = \sqrt{\dfrac{1 - \cos 315^0}{2}} = \sqrt{(1 - \sqrt{3}/2)/2} =$

$\sqrt{(2 - \sqrt{3})/4} = \sqrt{2 - \sqrt{3}}/2$

(c) $\tan \dfrac{\Pi}{8} = \dfrac{1 - \cos \dfrac{\Pi}{4}}{\sin \dfrac{\Pi}{4}} = \dfrac{\dfrac{2 - \sqrt{2}}{2}}{\sqrt{2}/2} = \sqrt{2} - 1$

13) $4 \sin \dfrac{x}{2} \cos \dfrac{x}{2} = 2 \left(2 \sin \dfrac{x}{2} \cos \dfrac{x}{2}\right) = 2 \left(\sin 2 \cdot \dfrac{x}{2}\right) = 2 \sin x$

16) $\csc 2u = \dfrac{1}{\sin 2u} = \dfrac{1}{2 \sin u \cos u} = \tfrac{1}{2} \csc u \sec u$

19) $\cos 4\theta = \cos (2 \cdot 2\theta) = 2 \cos^2 2\theta - 1 = 2 (2 \cos^2 \theta - 1)^2 - 1 =$
    $2 (4 \cos^4 \theta - 4 \cos^2 \theta + 1) - 1 = 8 \cos^4 \theta - 8 \cos^2 \theta + 1$

22) $\cos^4 x - \sin^4 x = (\cos^2 x + \sin^2 x)(\cos^2 x - \sin^2 x) =$
    $1 (\cos^2 x - \sin^2 x) = \cos 2x$

25) $2 \sin^2 2t = 2 (2 \sin t \cos t)^2 = 8 \sin^2 t \cos^2 t = 8 \cos^2 t (1 - \cos^2 t) =$
    $8 \cos^2 t - 8 \cos^4 t;$ Using the result of Exercise 19,
    $2 \sin^2 2t + \cos 4t = (8 \cos^2 t - 8 \cos^4 t) + (8 \cos^4 t - 8 \cos^2 t + 1) =$
    $8 \cos^2 t - 8 \cos^2 t + 8 \cos^4 t - 8 \cos^4 t + 1 = 1$

28) $\dfrac{1 + \sin 2v + \cos 2v}{1 + \sin 2v - \cos 2v} = \dfrac{1 + 2 \sin v \cos v + 2 \cos^2 v - 1}{1 + 2 \sin v \cos v - (1 - 2 \sin^2 v)} =$

$\dfrac{2 \cos^2 v + 2 \sin v \cos v}{2 \sin^2 v + 2 \sin v \cos v} = \dfrac{2 \cos v (\cos v + \sin v)}{2 \sin v (\sin v + \cos v)} = \cot v$

31) $\sin 2t + \sin t = 0 \Rightarrow 2 \sin t \cos t + \sin t = 0 \Rightarrow$

   $\sin t (2 \cos t + 1) = 0 \Rightarrow \sin t = 0$ or $\cos t = -\frac{1}{2} \Rightarrow$

   $t = 0, \Pi$ or $\frac{2\Pi}{3}, \frac{4\Pi}{3}$; $\quad 0°, 180°, 120°, 240°$

34) $\cos 2\theta - \tan \theta = 1 \Rightarrow 1 - 2 \sin^2 \theta - \dfrac{\sin \theta}{\cos \theta} = 1 \Rightarrow$

   $2 \sin^2 \theta \cos \theta + \sin \theta = 0 \Rightarrow \sin \theta (2 \sin \theta \cos \theta + 1) = 0 \Rightarrow$

   $\sin \theta = 0$ or $\sin 2\theta = -1 \Rightarrow \theta = 0, \Pi$ or $2\theta = \dfrac{3\Pi}{2}, \dfrac{7\Pi}{2} \Rightarrow$

   { if $0 \le \theta < 2\Pi$, then $0 \le 2\theta < 4\Pi$; $\therefore$ use 2 revolutions }

   $\theta = 0, \Pi, \dfrac{3\Pi}{4}, \dfrac{7\Pi}{4}$; $\quad 0°, 180°, 135°, 315°$

37) $\sin \dfrac{u}{2} + \cos u = 1 \Rightarrow \left[\sin \dfrac{u}{2}\right]^2 = (1 - \cos u)^2 \Rightarrow \dfrac{1 - \cos u}{2} =$

   $1 - 2 \cos u + \cos^2 u \Rightarrow 2 \cos^2 u - 3 \cos u + 1 = 0 \Rightarrow$

   $(2 \cos u - 1)(\cos u - 1) = 0 \Rightarrow \cos u = \frac{1}{2}, 1 \Rightarrow$

   $u = \dfrac{\Pi}{3}, \dfrac{5\Pi}{3}, 0$; $\quad 60°, 300°, 0°$ { All solutions check.}

40) $\sqrt{8^2 + 15^2} = 17$; Now $\sin v = \frac{15}{17}$ and $\cos v = \frac{8}{17} \Rightarrow v \simeq 1.08$

   radians or $\simeq 62°$; $\quad 8 \sin u + 15 \cos u \simeq 17 \sin (u + 1.08)$

43) (a) Example 2 in the text shows that $\cos 3x = 4 \cos^3 x - 3 \cos x$;

   $\cos 3x - 3 \cos x = 0 \Rightarrow (4 \cos^3 x - 3 \cos x) - 3 \cos x = 0 \Rightarrow$

   $2 \cos x (2 \cos^2 x - 3) = 0 \Rightarrow \cos x = 0, \pm\sqrt{1.5} \Rightarrow$

   $x = -\dfrac{3\Pi}{2}, -\dfrac{\Pi}{2}, \dfrac{\Pi}{2}, \dfrac{3\Pi}{2}$ { $\cos x \ne \sqrt{1.5}$ since $\sqrt{1.5} > 1$ }

   (b) Exercise 17 in the text shows that $\sin 3x = 3 \sin x - 4 \sin^3 x$;

   $-3 \sin 3x + 3 \sin x = 0 \Rightarrow \sin 3x - \sin x = 0 \Rightarrow$

   $(3 \sin x - 4 \sin^3 x) - \sin x = 0 \Rightarrow 4 \sin^3 x - 2 \sin x = 0 \Rightarrow$

Exercises 6.4

$$2 \sin x \, (2 \sin^2 x - 1) = 0 \implies \sin x = 0, \pm \frac{\sqrt{2}}{2} \implies$$

$$x = 0, \pm \Pi, \pm 2\Pi, \pm \frac{\Pi}{4}, \pm \frac{3\Pi}{4}, \pm \frac{5\Pi}{4}, \pm \frac{7\Pi}{4}$$

46) From Example 8 the area of a cross section is $\frac{1}{2} \left(\frac{1}{2}\right)^2 \sin \theta = \frac{1}{8} \sin \theta$. The volume is $20 \left(\frac{1}{8} \sin \theta\right) = \frac{5}{2} \sin \theta$; $\frac{5}{2} \sin \theta = 2 \implies \sin \theta = \frac{4}{5} \implies \theta \simeq 53.13^0$

Exercises 6.5

For this section, reference will be made to the product formulas as P1-P4 and the factoring formulas as F1-F4 in the order they appear in the text.

1) $2 \sin 9\theta \cos 3\theta = $ (P1) $\sin (9\theta + 3\theta) + \sin (9\theta - 3\theta) = \sin 12\theta + \sin 6\theta$

4) $\sin (-4x) \cos 8x = -\frac{1}{2} (2 \sin 4x \cos 8x) = $ (P1)
$-\frac{1}{2} [\sin (4x + 8x) + \sin (4x - 8x)] = -\frac{1}{2} \sin 12x - \frac{1}{2} \sin (-4x) = \frac{1}{2} \sin 4x - \frac{1}{2} \sin 12x$

7) $3 \cos x \sin 2x = \frac{3}{2} (2 \cos x \sin 2x) = $ (P2)
$\frac{3}{2} [\sin (x + 2x) - \sin (x - 2x)] = \frac{3}{2} \sin 3x + \frac{3}{2} \sin x$

10) $\sin 4\theta - \sin 8\theta = $ (F2) $2 \cos \dfrac{4\theta + 8\theta}{2} \sin \dfrac{4\theta - 8\theta}{2} =$
$-2 \cos 6\theta \sin 2\theta$

13) $\sin 3t - \sin 7t = $ (F2) $2 \cos \dfrac{3t + 7t}{2} \sin \dfrac{3t - 7t}{2} =$
$-2 \cos 5t \sin 2t$

16) $\sin 8t + \sin 2t = $ (F1) $2 \sin \dfrac{8t + 2t}{2} \cos \dfrac{8t - 2t}{2} =$
$2 \sin 5t \cos 3t$

19) $\dfrac{\sin u + \sin v}{\cos u + \cos v} = \dfrac{\text{(F1) } 2 \sin \dfrac{u + v}{2} \cos \dfrac{u - v}{2}}{\text{(F3) } 2 \cos \dfrac{u + v}{2} \cos \dfrac{u - v}{2}} = \tan \dfrac{u + v}{2}$

22) $\dfrac{\cos u - \cos v}{\cos u + \cos v} = \dfrac{\text{(F4) } 2 \sin \frac{v+u}{2} \sin \frac{v-u}{2}}{\text{(F3) } 2 \cos \frac{u+v}{2} \cos \frac{u-v}{2}} = \tan \dfrac{u+v}{2} \tan \dfrac{v-u}{2}$

25) $\sin ax \cos bx = \frac{1}{2} (2 \sin ax \cos bx) = \text{(P1)}$

$\frac{1}{2} [\sin (a+b) x + \sin (a-b) x] = \frac{1}{2} \sin (a+b) x + \frac{1}{2} \sin (a-b) x$

28) $\sin t + \sin 3t = \sin 2t \;\Rightarrow\; \text{(F1) } 2 \sin 2t \cos t = \sin 2t \;\Rightarrow\;$

$\sin 2t \, (2 \cos t - 1) = 0 \;\Rightarrow\; 2 \sin t \cos t \, (2 \cos t - 1) = 0 \;\Rightarrow\;$

$\sin t = 0$ or $\cos t = 0, \frac{1}{2} \;\Rightarrow\; t = \Pi n$ or $\frac{\Pi}{2} + \Pi n, \frac{\Pi}{3} + \Pi n, \frac{5\Pi}{3} + \Pi n$

$\{ \Pi n$ and $\frac{\Pi}{2} + \Pi n$ can be combined into $\frac{\Pi}{2} n \}$

31) $\sin (\theta + \Pi) = \sin \theta \cos \Pi + \cos \theta \sin \Pi = -\sin \theta + 0 = -\sin \theta$

34) $\cos (\theta - 3\Pi) = \cos \theta \cos 3\Pi + \sin \theta \sin 3\Pi = -\cos \theta + 0 = -\cos \theta$

37) $\cos x + \cos 3x = 0 \;\Rightarrow\; \text{(F3) } 2 \cos 2x \cos x = 0 \;\Rightarrow\; \cos 2x = 0$ or

$\cos x = 0 \;\Rightarrow\; 2x = \frac{\Pi}{2} + \Pi n$ or $x = \frac{\Pi}{2} + \Pi n \;\Rightarrow\; x = \frac{\Pi}{4} + \frac{\Pi}{2} n$ or

$x = \frac{\Pi}{2} + \Pi n \;\Rightarrow\; x = \frac{\Pi}{4}, \frac{3\Pi}{4}, \frac{5\Pi}{4}, \frac{7\Pi}{4}, \frac{\Pi}{2}, \frac{3\Pi}{2}$ for $0 \leq x \leq 2\Pi$

40) $\cos 4x - \cos x = 0 \;\Rightarrow\; \text{(F4) } -2 \sin \frac{5x}{2} \sin \frac{3x}{2} = 0 \;\Rightarrow\; \sin \frac{5x}{2} = 0$ or

$\sin \frac{3x}{2} = 0 \;\Rightarrow\; \frac{5x}{2} = \Pi n$ or $\frac{3x}{2} = \Pi n \;\Rightarrow\; x = \frac{2\Pi}{5} n$ or $\frac{2\Pi}{3} n;$

$x = 0, \pm \frac{2\Pi}{5}, \pm \frac{4\Pi}{5}, \pm \frac{6\Pi}{5}, \pm \frac{8\Pi}{5}, \pm 2\Pi, \pm \frac{2\Pi}{3}, \pm \frac{4\Pi}{3}$

Exercises 6.6

1) (a) $\sin^{-1} \frac{1}{2} = \Pi/6$ since $\sin (\Pi/6) = \frac{1}{2}$ and $-\Pi/2 \leq \Pi/6 \leq \Pi/2$

(b) $\cos^{-1} (\sqrt{3}/2) = \Pi/6$ since $\cos (\Pi/6) = \sqrt{3}/2$ and $0 \leq \Pi/6 \leq \Pi$

4) (a) $\cos^{-1} (\sqrt{2}/2) = \Pi/4$ since $\cos (\Pi/4) = \sqrt{2}/2$ and $0 \leq \Pi/4 \leq \Pi$

(b) $\cos^{-1} (-\sqrt{2}/2) = \frac{3\Pi}{4}$ since $\cos \left(\frac{3\Pi}{4} \right) = -\sqrt{2}/2$ and $0 \leq \frac{3\Pi}{4} \leq \Pi$

Exercises 6.6

7) (a) $\tan^{-1}(\sqrt{3}) = \Pi/3$ since $\tan(\Pi/3) = \sqrt{3}$ and $-\Pi/2 < \Pi/3 < \Pi/2$

(b) $\arctan(-\sqrt{3}) = -\dfrac{\Pi}{3}$ since $\tan\left(-\dfrac{\Pi}{3}\right) = -\sqrt{3}$ and $-\dfrac{\Pi}{2} < -\dfrac{\Pi}{3} < \dfrac{\Pi}{2}$

10) (a) 0.2007　　　　　　　　　(b) 0.20074559

13) (a) 1.1403　　　　　　　　　(b) 1.1402832

16) $\cos(\sin^{-1} 0) = \cos 0 = 1$

19) $\arcsin\left[\sin\dfrac{5\Pi}{4}\right] = \arcsin\left(-\dfrac{\sqrt{2}}{2}\right) = -\dfrac{\Pi}{4}$

22) $\arcsin\dfrac{1}{2} = \dfrac{\Pi}{6}$ and $\arccos 0 = \dfrac{\Pi}{2}$; $\sin\left(\dfrac{\Pi}{6} + \dfrac{\Pi}{2}\right) = \sin\dfrac{2\Pi}{3} = \dfrac{\sqrt{3}}{2}$

25) Let $\alpha = \arccos\left(-\dfrac{3}{5}\right)$. Now $\cos\alpha = -\dfrac{3}{5}$ and $\sin\alpha = \dfrac{4}{5}$. See Figure
　　25. $\sin[2\arccos(-\dfrac{3}{5})] = \sin 2\alpha = 2\sin\alpha\cos\alpha = 2\cdot\dfrac{4}{5}\cdot\dfrac{-3}{5} = \dfrac{-24}{25}$

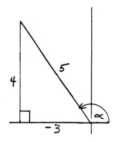

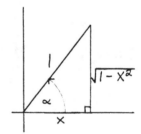

Figure 25　　　　　　　　　　　　　　Figure 28

28) Sketch a right triangle with legs x and $\sqrt{1-x^2}$ and hypotenuse 1
　　with angle α so that $\cos\alpha = \dfrac{x}{1}$. See Figure 28. Then

$$\tan(\arccos x) = \tan\alpha = \dfrac{\sqrt{1-x^2}}{x}.$$

31) Let $\alpha = \sin^{-1} x$ and $\beta = \cos^{-1} x$. Now $-\dfrac{\Pi}{2} \le \alpha \le \dfrac{\Pi}{2}$ and $0 \le \beta \le \Pi$

　　so $-\dfrac{\Pi}{2} \le \alpha + \beta \le \dfrac{3\Pi}{2}$. $\sin(\alpha + \beta) = \sin\alpha\cos\beta + \cos\alpha\sin\beta =$

　　$x\cdot x + \sqrt{1-x^2}\cdot\sqrt{1-x^2} = x^2 + (1 - x^2) = 1$. Since the only number in $\left[-\dfrac{\Pi}{2}, \dfrac{3\Pi}{2}\right]$ whose sine is 1 is $\dfrac{\Pi}{2}$, it follows that

$\alpha + \beta = \dfrac{\Pi}{2}$.

34) Let $\alpha = \cos^{-1} x$ and $\beta = \cos^{-1}(2x^2 - 1)$. Since $0 \le x \le 1$, $0 \le \alpha \le \dfrac{\Pi}{2}$

$(0 \le 2\alpha \le \Pi)$ and $0 \le \beta \le \Pi$. Now $\cos 2\alpha = \cos^2 \alpha - \sin^2 \alpha =$

$x^2 - (1 - x^2) = 2x^2 - 1$ and $\cos \beta = 2x^2 - 1$. The cosine function is

one-to-one on $[0, \Pi]$, so $2\alpha = \beta$.

37) Let $\alpha = \sin^{-1} x$ and $\beta = \tan^{-1} \dfrac{x}{\sqrt{1 - x^2}}$. $-\dfrac{\Pi}{2} < \alpha, \beta < \dfrac{\Pi}{2}$

Now $\sin \alpha = x$ and $\sin \beta = x$. The sketch is like Figure 28 in this

section with the leg labelings reversed. The sine function is one-to-

one on $[-\Pi/2, \Pi/2]$, so $\alpha = \beta$.

40) $\text{Sec}^{-1} x = y$ iff $\sec y = x$ where $y \in [0, \Pi/2) \cup [\Pi, 3\Pi/2)$.

43) The "period" of $y = \sin^{-1} x$ is 2 units. $\{-1 \text{ to } 1\}$ The period of

$y = \sin^{-1} 2x$ is $\dfrac{2}{2} = 1$ unit. $\{-\dfrac{1}{2} \text{ to } \dfrac{1}{2}\}$ See Figure 43.

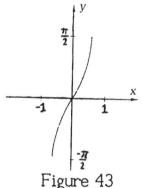

Figure 43

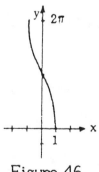

Figure 46

46) The range of $y = \cos^{-1} x$ is $[0, \Pi]$. The range of $y = 2 \cos^{-1} x$ is

$[0, 2\Pi]$. y-intercept : Π; See Figure 46.

49) Refer to Figure 6.10. The horizontal asymptotes for $y = \tan^{-1} x$ are

$y = -\dfrac{\Pi}{2}$ and $y = \dfrac{\Pi}{2}$. The horizontal asymptotes for $y = 2 + \tan^{-1} x$

are $y = 2 \pm \frac{\Pi}{2}$; y-intercept : 2; See Figure 49.

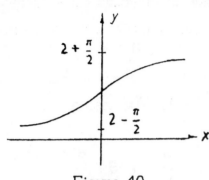

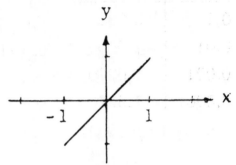

Figure 49   Figure 52

52) $y = \sin(\sin^{-1} x) = x$, which is the equation of the line with slope 1 and y-intercept 0. Remember that the domain of $y = \sin^{-1} x$ is $[-1, 1]$; this is why the graph is only a segment of $y = x$. See Figure 52.

55) (a) The equation can be thought of as a quadratic equation in $\tan t$.

$$\tan t = \frac{-9 \pm \sqrt{81 - 24}}{4} \Rightarrow t = \arctan\left(\frac{-9 \pm \sqrt{57}}{4}\right)$$

(b) -1.3337, -0.3478

58) (a) The equation can be thought of as a quadratic equation in $\tan^2 \theta$.

$$\tan^2 \theta = \frac{19 \pm \sqrt{361 - 24}}{6} \Rightarrow \theta = \arctan\left(\pm\sqrt{(19 \pm \sqrt{337})/6}\right)$$

(b) ± 0.3162, ± 1.1896

Calculator Exercises 6.6

1) $\sin^{-1}(\sin 2) \simeq 1.1416 \neq 2$; 2 is not the answer because 2 is outside the range of the arcsin function. $\cos^{-1}(\cos 2) = 2$ because the arccos function is defined to have a range from 0 to Π.

4) No, the domain of the arctan is all reals so the arctangent of any number is defined.

7)

x	csc x tan⁻¹ x
0.1	0.9983
0.01	1.0000
0.001	1.0000
0.0001	1.0000

From the results in the table, we see that as x gets close to 0 through values greater than 0, $f(x) = \csc x \tan^{-1} x$ gets very close to 1.

1) $(\cot^2 x + 1)(1 - \cos^2 x) = (\csc^2 x)(\sin^2 x) = 1$

4) $(\tan x + \cot x)^2 = \left(\dfrac{\sin x}{\cos x} + \dfrac{\cos x}{\sin x}\right)^2 = \left(\dfrac{\sin^2 x + \cos^2 x}{\cos x \sin x}\right)^2 =$

$\dfrac{1}{\cos^2 x \sin^2 x} = \sec^2 x \csc^2 x$

7) $\tan 2u = \dfrac{2 \tan u}{1 - \tan^2 u} \cdot \dfrac{\cot^2 u}{\cot^2 u} = \dfrac{2 \cot u}{\cot^2 u - 1} = \dfrac{2 \cot u}{\csc^2 u - 1 - 1} = \dfrac{2 \cot u}{\csc^2 u - 2}$

10) $\text{LHS} = \dfrac{\sin u + \sin v}{\csc u + \csc v} = \dfrac{\sin u + \sin v}{\dfrac{1}{\sin u} + \dfrac{1}{\sin v}} = \dfrac{\sin u + \sin v}{\dfrac{\sin u + \sin v}{\sin u \sin v}} = \sin u \sin v$

$\text{RHS} = \dfrac{1 - \sin u \sin v}{-1 + \csc u \csc v} = \dfrac{1 - \sin u \sin v}{\dfrac{1 - \sin u \sin v}{\sin u \sin v}} = \sin u \sin v$

Since the Left Hand Side and Right Hand Side equal the same expression and the steps are reversible, the identity is verified.

13) $\frac{1}{4} \sin 4\beta = \frac{1}{4}(2 \sin 2\beta \cos 2\beta) = \frac{1}{2}(2 \sin \beta \cos \beta (\cos^2 \beta - \sin^2 \beta)) = \sin \beta \cos^3 \beta - \cos \beta \sin^3 \beta$

16) Let $\alpha = \arctan x$ and $\beta = \arctan \dfrac{2x}{1 - x^2}$. Since $|x| \le 1$,

$-\dfrac{\Pi}{2} \le \alpha \le \dfrac{\Pi}{2}$ and $-\dfrac{\Pi}{2} \le \beta \le \dfrac{\Pi}{2}$. Now $\tan \alpha = x$ and $\tan \frac{1}{2}\beta =$

Exercises 6.7

$$\frac{1 - \cos \beta}{\sin \beta} = \frac{1 - \dfrac{1 - x^2}{1 + x^2}}{\dfrac{2x}{1 + x^2}} = \frac{\dfrac{2x^2}{1 + x^2}}{\dfrac{2x}{1 + x^2}} = x. \quad \text{To obtain the } 1 + x^2$$

consider a right triangle with legs $1 - x^2$ and $2x$. Then the

hypotenuse is $\sqrt{(1 - x^2)^2 + (2x)^2} = \sqrt{x^4 + 2x^2 + 1} = \sqrt{(x^2 + 1)^2} =$

$x^2 + 1$. The tangent function is 1-1 on $\left[-\dfrac{\Pi}{2}, \dfrac{\Pi}{2}\right]$, so $\alpha = \dfrac{1}{2}\beta$.

19) $\sin \theta = \tan \theta \implies \sin \theta = \dfrac{\sin \theta}{\cos \theta} \implies \dfrac{\sin \theta}{\cos \theta} - \sin \theta = 0 \implies$

$\sin \theta \left(\dfrac{1}{\cos \theta} - 1\right) = 0 \implies \sin \theta \left(\dfrac{1 - \cos \theta}{\cos \theta}\right) = 0 \implies$

$\implies \sin \theta = 0$ or $\cos \theta = 1 \implies \theta = 0, \Pi$ or $\theta = 0; \quad 0^0, 180^0$

22) $\cos x \cot^2 x = \cos x \implies \cos x (\cot^2 x - 1) = 0 \implies$

$\cos x = 0$ or $\cot x = \pm 1 \implies x = \dfrac{\Pi}{2}, \dfrac{3\Pi}{2}, \dfrac{\Pi}{4}, \dfrac{5\Pi}{4}, \dfrac{3\Pi}{4}, \dfrac{7\Pi}{4};$

$90^0, 270^0, 45^0, 225^0, 135^0, 315^0$

25) $2 \sec u \sin u + 2 = 4 \sin u + \sec u \implies$

$2 \sec u \sin u - 4 \sin u - \sec u + 2 = 0 \implies$

$2 \sin u (\sec u - 2) - 1 (\sec u - 2) = 0 \implies$

$(2 \sin u - 1)(\sec u - 2) = 0 \implies \sin u = \tfrac{1}{2}$ or $\sec u = 2 \implies$

$u = \dfrac{\Pi}{6}, \dfrac{5\Pi}{6}, \dfrac{\Pi}{3}, \dfrac{5\Pi}{3}; \quad 30^0, 150^0, 60^0, 300^0$

28) $\sec 2x \csc 2x = 2 \csc 2x \implies \csc 2x (\sec 2x - 2) = 0 \implies$

$\sec 2x = 2 \ \{\csc 2x \neq 0\} \implies 2x = \dfrac{\Pi}{3}, \dfrac{5\Pi}{3}, \dfrac{7\Pi}{3}, \dfrac{11\Pi}{3} \implies$

$x = \dfrac{\Pi}{6}, \dfrac{5\Pi}{6}, \dfrac{7\Pi}{6}, \dfrac{11\Pi}{6}; \quad 30^0, 150^0, 210^0, 330^0$

31) $\sin 195° = \sin (135° + 60°) = \sin 135° \cos 60° + \cos 135° \sin 60° =$

$\dfrac{\sqrt{2}}{2} \cdot \dfrac{1}{2} + \dfrac{-\sqrt{2}}{2} \cdot \dfrac{\sqrt{3}}{2} = \dfrac{\sqrt{2} - \sqrt{6}}{4}$

34) $\cos (\theta + \phi) = \cos \theta \cos \phi - \sin \theta \sin \phi = \frac{4}{5} \cdot \frac{8}{17} - \frac{3}{5} \cdot \frac{15}{17} = -\frac{13}{85}$

37) $\sin 2\phi = 2 \sin \phi \cos \phi = 2 \cdot \frac{15}{17} \cdot \frac{8}{17} = \frac{240}{289}$

40) $\sin \dfrac{\theta}{2} = \sqrt{\dfrac{1 - \cos \theta}{2}} = \sqrt{\dfrac{1 - 4/5}{2}} = \sqrt{\dfrac{1}{10}} = \dfrac{\sqrt{10}}{10}$

43) (a) $\sin 7t \sin 4t = \frac{1}{2} (2 \sin 7t \sin 4t) =$ (P4)

$\quad\quad \frac{1}{2} [\cos (7t - 4t) - \cos (7t + 4t)] = \frac{1}{2} \cos 3t - \frac{1}{2} \cos 11t$

(b) $\cos \dfrac{u}{4} \cos \left(- \dfrac{u}{6}\right) = \dfrac{1}{2}\left(2 \cos \dfrac{u}{4} \cos \dfrac{u}{6}\right) =$ (P3)

$\quad\quad \dfrac{1}{2}\left[\cos \left(\dfrac{u}{4} + \dfrac{u}{6}\right) + \cos \left(\dfrac{u}{4} - \dfrac{u}{6}\right)\right] = \dfrac{1}{2} \cos \dfrac{5u}{12} + \dfrac{1}{2} \cos \dfrac{u}{12}$

(c) $6 \cos 5x \sin 3x = 3 (2 \cos 5x \sin 3x) =$ (P2)

$\quad\quad 3 [\sin (5x + 3x) - \sin (5x - 3x)] = 3 \sin 8x - 3 \sin 2x$

46) $\sin^{-1} \left(- \dfrac{\sqrt{2}}{2}\right) = -\dfrac{\Pi}{4}$ since $\sin \left(-\dfrac{\Pi}{4}\right) = -\dfrac{\sqrt{2}}{2}$ and $-\dfrac{\Pi}{4}$ is in the range

of the arcsin function.

49) $\sin \left[\arccos \left(-\dfrac{\sqrt{3}}{2}\right)\right] = \sin \left(\dfrac{5\Pi}{6}\right) = \frac{1}{2}$

52) $\sin (\sin^{-1} \frac{2}{3}) = \frac{2}{3}$

55) The range of $y = 4 \sin^{-1} x$ is
$[-2\Pi, 2\Pi]$. The 4 has the
effect of vertically stretching
{ 4 times } the graph of
$y = \sin^{-1} x$. See Figure 55.

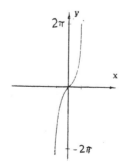

Figure 55

58) $\cos x - \cos 2x + \cos 3x = 0 \implies (\cos x + \cos 3x) - \cos 2x = 0 \implies$

(F3) $2 \cos \dfrac{x + 3x}{2} \cos \dfrac{x - 3x}{2} - \cos 2x = 0 \implies$

$2 \cos 2x \cos x - \cos 2x = 0 \implies \cos 2x (2 \cos x - 1) = 0 \implies$

$\cos 2x = 0$ or $\cos x = \tfrac{1}{2} \implies 2x = \dfrac{\Pi}{2} + \Pi n \left(x = \dfrac{\Pi}{4} + \dfrac{\Pi}{2}n \right)$ or

$x = \dfrac{\Pi}{3} + 2\Pi n, \ \dfrac{5\Pi}{3} + 2\Pi n$

1) $\beta = 180^0 - \alpha - \gamma = 62^0$;

$$\frac{b}{\sin\beta} = \frac{a}{\sin\alpha} \Rightarrow b = \frac{a\sin\beta}{\sin\alpha} = \frac{10.5\sin 62^0}{\sin 41^0} \approx 14.1;$$

$$\frac{c}{\sin\gamma} = \frac{a}{\sin\alpha} \Rightarrow c = \frac{a\sin\gamma}{\sin\alpha} = \frac{10.5\sin 77^0}{\sin 41^0} \approx 15.6$$

4) $\beta = 180^0 - \alpha - \gamma = 76^0 30'$;

$$\frac{a}{\sin\alpha} = \frac{b}{\sin\beta} \Rightarrow a = \frac{b\sin\alpha}{\sin\beta} = \frac{19.7\sin 42^0 10'}{\sin 76^0 30'} \approx 13.6;$$

$$\frac{c}{\sin\gamma} = \frac{b}{\sin\beta} \Rightarrow c = \frac{b\sin\gamma}{\sin\beta} = \frac{19.7\sin 61^0 20'}{\sin 76^0 30'} \approx 17.8$$

7) $\dfrac{\sin\beta}{b} = \dfrac{\sin\alpha}{a} \Rightarrow \beta = \sin^{-1}\left(\dfrac{b\sin\alpha}{a}\right) \approx \sin^{-1}(0.8053) \approx 53^0 40'$ or

$126^0 20'$. {rounded to the nearest 10 minutes} Reject $126^0 20'$
because then $\alpha + \beta \geq 180^0$. $\gamma = 180^0 - \alpha - \beta \approx 61^0 10'$;

$$\frac{c}{\sin\gamma} = \frac{a}{\sin\alpha} \Rightarrow c = \frac{a\sin\gamma}{\sin\alpha} \approx 20.6$$

10) $\dfrac{\sin\gamma}{c} = \dfrac{\sin\alpha}{a} \Rightarrow \gamma = \sin^{-1}\left(\dfrac{c\sin\alpha}{a}\right) \approx \sin^{-1}(0.8676) \approx 60^0 10'$ or

$119^0 50'$. {rounded to the nearest 10 minutes} There are two
triangles possible since in either case $\alpha + \gamma < 180^0$.
$\beta = 180^0 - \alpha - \gamma \approx 92^0 20'$ or $32^0 40'$;

$$\frac{b}{\sin\beta} = \frac{a}{\sin\alpha} \Rightarrow b = \frac{a\sin\beta}{\sin\alpha} \approx 60.8 \text{ or } 32.8$$

13) $\angle ABC = 180^0 - 54^0 10' - 63^0 20' = 62^0 30'$;

$$\frac{\overline{AB}}{\sin 54^0 10'} = \frac{240}{\sin 62^0 30'} \Rightarrow \overline{AB} \approx 219.36 \text{ yards}$$

Exercises 7.1

16) See Figure 16. $\angle BAC = 57^0 - 15^0 = 42^0$;

$\angle ABC = 90^0 + 15^0 = 105^0$; $\angle BCA = 180^0 - 42^0 - 105^0 = 33^0$;

$$\frac{\overline{BC}}{\sin 42^0} = \frac{75}{\sin 33^0} \Rightarrow \overline{BC} \approx 92.14 \text{ feet}$$

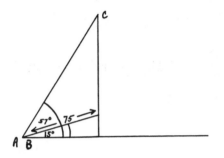

Figure 16   Figure 19

19) See Figure 19. $\angle FAB = 90^0 - 27^0 10' = 62^0 50'$;

$\angle FBA = 90^0 - 52^0 40' = 37^0 20'$;

$\angle AFB = 180^0 - 62^0 50' - 37^0 20' = 79^0 50'$;

$$\frac{\overline{AF}}{\sin 37^0 20'} = \frac{6}{\sin 79^0 50'} \Rightarrow \overline{AF} \approx 3.70 \text{ mi.}$$

$$\frac{\overline{BF}}{\sin 62^0 50'} = \frac{6}{\sin 79^0 50'} \Rightarrow \overline{BF} \approx 5.42 \text{ mi.}$$

22) (a) Label the base of the tower A, the top of the tower B, and the observation point C. Let $\alpha = \angle BAC$ and $\beta = \angle ABC$.

$$\frac{\sin \beta}{150} = \frac{\sin 53.3^0}{179} \Rightarrow \beta \approx 42.2^0;$$

$\alpha \approx 180^0 - 42.2^0 - 53.3^0 = 84.5^0$; $\theta = 90^0 - \alpha = 5.5^0$

(b) $d = 179 \sin \theta \approx 17.2 \text{ feet}$

25) In the triangle that forms the base, let $\alpha = 103^0$, $\beta = 52^0$, and

$\gamma = 180^0 - \alpha - \beta = 25^0$. $\dfrac{a}{\sin \alpha} = \dfrac{12}{\sin \gamma} \Rightarrow a \approx 27.7 \text{ units.}$ Now

$\tan 34^0 = \dfrac{h}{a}$ $\Rightarrow$ $h \simeq 18.7$ units. The volume of a prism is $\frac{1}{3} Bh$, so we need the area of the base, B. Draw a line from α that is perpendicular to a and call it d. $d = 12 \sin 52^0 \simeq 9.5$ units. The area of the base is $\frac{1}{2} ad$. Now the volume is $\frac{1}{3} Bh = \frac{1}{3} (\frac{1}{2} ad) h = 288 \sin 52^0 \sin^2 103^0 \tan 34^0 \csc^2 25^0 \simeq 814$ cubic units.

1) $\alpha = 180^0 - \beta - \gamma = 119.7^0$;

$\dfrac{a}{\sin \alpha} = \dfrac{b}{\sin \beta}$ $\Rightarrow$ $a = \dfrac{b \sin \alpha}{\sin \beta} \simeq 371.5$;

$\dfrac{c}{\sin \gamma} = \dfrac{b}{\sin \beta}$ $\Rightarrow$ $c = \dfrac{b \sin \gamma}{\sin \beta} \simeq 243.5$

4) $\dfrac{\sin \alpha}{a} = \dfrac{\sin \gamma}{c}$ $\Rightarrow$ $\alpha = \sin^{-1}\left(\dfrac{a \sin \gamma}{c}\right) \simeq \sin^{-1} (0.9915) \simeq 82.54^0$ or

97.46^0. There are two triangles possible since in either case $\alpha + \gamma < 180^0$. $\beta = 180^0 - \alpha - \gamma \simeq 49.72^0$ or 34.80^0;

$\dfrac{b}{\sin \beta} = \dfrac{c}{\sin \gamma}$ $\Rightarrow$ $b = \dfrac{c \sin \beta}{\sin \gamma} \simeq 100.85$ or 75.45

Using the 3 forms of the Law of Cosines from the text, we derive the following formulas for the 6 parts of a triangle :

1) $a = \sqrt{b^2 + c^2 - 2bc \cos \alpha}$

4) $\alpha = \cos^{-1}\left(\dfrac{b^2 + c^2 - a^2}{2bc}\right)$

2) $b = \sqrt{a^2 + c^2 - 2ac \cos \beta}$

5) $\beta = \cos^{-1}\left(\dfrac{a^2 + c^2 - b^2}{2ac}\right)$

3) $c = \sqrt{a^2 + b^2 - 2ab \cos \gamma}$

6) $\gamma = \cos^{-1}\left(\dfrac{a^2 + b^2 - c^2}{2ab}\right)$

These formulas will be used without mention in the Exercise solutions.

Exercises 7.2

1) $a = \sqrt{b^2 + c^2 - 2bc \cos \alpha} = \sqrt{700} \simeq 26$;

$\beta = \cos^{-1}\left(\dfrac{a^2 + c^2 - b^2}{2ac}\right) \simeq \cos^{-1}(0.7559) \simeq 41°$;

$\gamma = 180° - \alpha - \beta \simeq 79°$

4) $b = \sqrt{a^2 + c^2 - 2ac \cos \beta} \simeq \sqrt{7086.74} \simeq 84.2$;

$\alpha = \cos^{-1}\left(\dfrac{b^2 + c^2 - a^2}{2bc}\right) \simeq \cos^{-1}(-0.1214) \simeq 97°$;

$\gamma = 180° - \alpha - \beta = 9°10'$

7) $\alpha = \cos^{-1}\left(\dfrac{b^2 + c^2 - a^2}{2bc}\right) = \cos^{-1}(0.875) \simeq 29°$;

$\beta = \cos^{-1}\left(\dfrac{a^2 + c^2 - b^2}{2ac}\right) = \cos^{-1}(0.6875) \simeq 47°$;

$\gamma = 180° - \alpha - \beta \simeq 104°$

10) $\alpha = \cos^{-1}\left(\dfrac{b^2 + c^2 - a^2}{2bc}\right) = \cos^{-1}(0.25) \simeq 75°30'$;

$\beta = \cos^{-1}\left(\dfrac{a^2 + c^2 - b^2}{2ac}\right) = \cos^{-1}(0.25) \simeq 75°30'$;

$\gamma = 180° - \alpha - \beta \simeq 29°$

13) 20 minutes $= \frac{1}{3}$ hour $\Rightarrow$ the cars
have travelled 20 miles $\{60 \cdot \frac{1}{3}\}$
and 15 miles $\{45 \cdot \frac{1}{3}\}$, respectively.
The distance apart, d, is given by
$d = \sqrt{20^2 + 15^2 - 2(20)(15) \cos 84°}$
$\simeq 24$ miles; See Figure 13.

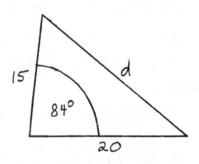

Figure 13

16) See Figure 16. The angle between the two paths is $40^0 + 25^0 = 65^0$.

$\overline{AB} = \sqrt{165^2 + 80^2 - 2\,(165)\,(80)\cos 65^0} \simeq 150$ miles;

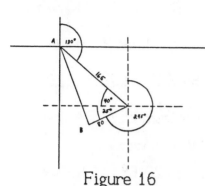

Figure 16   Figure 19

19) See Figure 19. Let α be the angle opposited the side of length 3.

$\alpha = \cos^{-1}\left[\dfrac{2^2 + 4^2 - 3^2}{2\,(2)\,(4)}\right] = \cos^{-1}(0.6875) \simeq 46^034'$;

Let β be the angle opposite the side of length 2.

$\beta = \cos^{-1}\left[\dfrac{3^2 + 4^2 - 2^2}{2\,(3)\,(4)}\right] = \cos^{-1}(0.875) \simeq 28^057'$;

Now $\overline{AB}$ is N$26^034'$E from point B. The direction of the third side is N$(26^034' + 28^057')$E, or N$55^031'$E.

22) The diagonal in the base is $\sqrt{6^2 + 8^2} = \sqrt{100} = 10$ inches;

The diagonal of the $6'' \times 4''$ side is $\sqrt{4^2 + 6^2} = \sqrt{52}$ inches;

The diagonal of the $8'' \times 4''$ side is $\sqrt{8^2 + 4^2} = \sqrt{80}$ inches. Now

$\theta = \cos^{-1}\left[\dfrac{100 + 52 - 80}{2\,(\sqrt{100}\,)\,(\sqrt{52}\,)}\right] \simeq \cos^{-1}(0.4992) \simeq 60^03'$.

25) (a) Call the 4 wing corners A, B, C, and D where A is the corner by angle ϕ and the others are labelled in a counterclockwise direction. In ΔDAC, $\angle CDA = 180^0 - 136^0 = 44^0$ and

$d = \overline{AC} = \sqrt{22.9^2 + 17.2^2 - 2\,(22.9)\,(17.2)\cos 44^0} \simeq 15.9$;

Exercises 7.2

Let $\alpha = \angle DAC$; Using the Law of Sines, $\dfrac{\sin \alpha}{22.9} = \dfrac{\sin 44^0}{d} \Rightarrow$

$\alpha = \sin^{-1}\left(\dfrac{22.9 \sin 44^0}{d}\right) \simeq 87.4^0$; Let $\beta = \angle CAB$; Using the

Law of Cosines, $5.7^2 = d^2 + 16^2 - 2\,(d)\,(16) \cos \beta \Rightarrow \beta =$

$\cos^{-1}\left(\dfrac{d^2 + 16^2 - 5.7^2}{2\,(d)\,(16)}\right) \simeq 20.6^0$; $\phi \simeq 180^0 - 87.4^0 - 20.6^0 =$

72^0; The area could be computed as the sum of $\triangle CDA$ and

$\triangle ABC$. Using triangles, we have A {Wing Area} =

$\frac{1}{2}\,(17.2)\,(22.9) \sin 44^0 + \frac{1}{2}\,(15.9)\,(16) \sin 20.6^0 \simeq 136.8 +$

$44.8 = 181.6$ square feet.

(b) $\angle CDA = 44^0$ so $\sin 44^0 = \dfrac{h}{22.9} \Rightarrow h \simeq 15.9$ is the length of one

wing (from body to tip). The wing span is $2h + 5.8$ or approximately 37.6 feet.

For Exercises 28, 31, and 34, A is measured in square units.

28) $A = \frac{1}{2} ab \sin \gamma = \frac{1}{2}\,(15)\,(10) \sin 45^0 = (37.5) \sqrt{2} \simeq 53.0$

31) Using Heron's Formula with $a = 2$, $b = 3$, and $c = 4$, we have

$s = \frac{1}{2}\,(a + b + c) = \frac{1}{2}\,(2 + 3 + 4) = 4.5$;

$A = \sqrt{s\,(s - a)\,(s - b)\,(s - c)} = \sqrt{(4.5)\,(2.5)\,(1.5)\,(0.5)} \simeq 2.9$

34) Using Heron's Formula with $a = 20$, $b = 20$, and $c = 10$, we have

$s = \frac{1}{2}\,(a + b + c) = \frac{1}{2}\,(20 + 20 + 10) = 25$;

$A = \sqrt{s\,(s - a)\,(s - b)\,(s - c)} = \sqrt{(25)\,(5)\,(5)\,(15)} \simeq 96.8$

Calculator Exercises 7.2

1) $a = \sqrt{b^2 + c^2 - 2bc \cos \alpha} \simeq \sqrt{1662.8} \simeq 40.8$;

$\beta = \cos^{-1}\left(\dfrac{a^2 + c^2 - b^2}{2ac}\right) \simeq \cos^{-1}(0.8919) \simeq 26.9^0$;

$\gamma = 180^0 - \alpha - \beta \simeq 104.8^0$

4) $\alpha = \cos^{-1}\left(\dfrac{b^2 + c^2 - a^2}{2bc}\right) \simeq \cos^{-1}(0.6541) \simeq 49.1°;$

$\beta = \cos^{-1}\left(\dfrac{a^2 + c^2 - b^2}{2ac}\right) \simeq \cos^{-1}(-0.5455) \simeq 123.1°;$

$\gamma = 180° - \alpha - \beta \simeq 7.8°$

1) $|3 - 4i| = \sqrt{3^2 + (-4)^2} = \sqrt{25} = 5$

4) $|1 - i| = \sqrt{1^2 + (-1)^2} = \sqrt{2}$

7) $|i^{500}| = |(i^4)^{125}| = |(1)^{125}| = |1 + 0i| = \sqrt{1^2 + 0^2} = 1$

10) $|-15 + 0i| = \sqrt{(-15)^2 + 0^2} = 15$

The geometric representations in Exercises 13, 16, and 19 are the following points :

13) $P(3, -5)$

16) $(1 + 2i)^2 = -3 + 4i$; $P(-3, 4)$

19) $(1 + i)^2 = 0 + 2i$; $P(0, 2)$

For Exercises 22-49, $\cos\theta + i\sin\theta$ will be represented by cis θ.

22) $r = 2$; $\tan\theta = \dfrac{1}{\sqrt{3}}$ and θ in QI $\Rightarrow$ $\theta = \dfrac{\Pi}{6}$; $2\,\text{cis}\,\dfrac{\Pi}{6}$

25) $r = 20$; $\tan\theta = \dfrac{-20}{0}$ (undefined) $\Rightarrow$ $\theta = \dfrac{3\Pi}{2}$; $20\,\text{cis}\,\dfrac{3\Pi}{2}$

28) $r = 6$; $\tan\theta = \dfrac{-6}{0}$ (undefined) $\Rightarrow$ $\theta = \dfrac{3\Pi}{2}$; $6\,\text{cis}\,\dfrac{3\Pi}{2}$

31) $r = \sqrt{5}$; $\tan\theta = \frac{1}{2}$ and θ in QI $\Rightarrow$ $\theta = \tan^{-1}\frac{1}{2}$; $\sqrt{5}\,\text{cis}\,(\tan^{-1}\frac{1}{2})$

34) $r = 10\sqrt{2}$; $\tan\theta = \dfrac{10}{-10} = -1$ and θ in QII $\Rightarrow$ $\theta = \dfrac{3\Pi}{4}$;
    $10\sqrt{2}\,\text{cis}\,\dfrac{3\Pi}{4}$

37) $r = 2$; $\tan\theta = \dfrac{-1}{\sqrt{3}}$ and θ in QIV $\Rightarrow$ $\theta = \dfrac{11\Pi}{6}$; $2\,\text{cis}\,\dfrac{11\Pi}{6}$

40) $r = 0$; θ could be any angle; $0 \text{ cis } \theta$

43) $z_1 = 4 \text{ cis } \dfrac{4\Pi}{3}$; $z_2 = 5 \text{ cis } \dfrac{\Pi}{2}$;

$z_1 z_2 = 4 \cdot 5 \text{ cis } \left(\dfrac{4\Pi}{3} + \dfrac{\Pi}{2} \right) = 20 \text{ cis } \dfrac{11\Pi}{6} = 10\sqrt{3} - 10i$;

$\dfrac{z_1}{z_2} = \dfrac{4}{5} \text{ cis } \left(\dfrac{4\Pi}{3} - \dfrac{\Pi}{2} \right) = \dfrac{4}{5} \text{ cis } \dfrac{5\Pi}{6} = \dfrac{-2\sqrt{3}}{5} + \dfrac{2}{5}i$

46) $z_1 = 2 \text{ cis } \dfrac{\Pi}{2}$; $z_2 = 3 \text{ cis } \dfrac{3\Pi}{2}$;

$z_1 z_2 = 2 \cdot 3 \text{ cis } \left(\dfrac{\Pi}{2} + \dfrac{3\Pi}{2} \right) = 6 \text{ cis } 2\Pi = 6 + 0i$;

$\dfrac{z_1}{z_2} = \dfrac{2}{3} \text{ cis } \left(\dfrac{\Pi}{2} - \dfrac{3\Pi}{2} \right) = \dfrac{2}{3} \text{ cis } (-\Pi) = -\dfrac{2}{3} + 0i$;

49) Let $z_1 = r_1 \text{ cis } \theta_1$, $z_2 = r_2 \text{ cis } \theta_2$, and $z_3 = r_3 \text{ cis } \theta_3$. Then $z_1 z_2 z_3 =$

$(z_1 z_2) z_3 = [r_1 r_2 \text{ cis } (\theta_1 + \theta_2)] \, [r_3 \text{ cis } \theta_3] =$

$(r_1 r_2) r_3 \text{ cis } ((\theta_1 + \theta_2) + \theta_3) = r_1 r_2 r_3 \text{ cis } (\theta_1 + \theta_2 + \theta_3)$.

The generalization is $z_1 z_2 \cdots z_n = r_1 r_2 \cdots r_n \text{ cis } \left(\theta_1 + \theta_2 + \cdots + \theta_n \right)$.

Exercises 7.4

In this section, it is assumed the reader can transform complex numbers to their trigonometric form. If this is not true, see Exercises 7.3

1) $(3 + 3i)^5 = \left[3\sqrt{2} \text{ cis } \dfrac{\Pi}{4} \right]^5 = (3\sqrt{2})^5 \text{ cis } \dfrac{5\Pi}{4} = 972\sqrt{2} \left(-\dfrac{\sqrt{2}}{2} - \dfrac{\sqrt{2}}{2} i \right)$

$= -972 - 972i$

4) $(-1 + i)^8 = \left[\sqrt{2} \text{ cis } \dfrac{3\Pi}{4} \right]^8 = (\sqrt{2})^8 \text{ cis } 6\Pi = 16 \text{ cis } 0 = 16 \, (1 + 0i)$

$= 16$

7) $\left(-\dfrac{\sqrt{2}}{2} + \dfrac{\sqrt{2}}{2} i \right)^{15} = \left[1 \text{ cis } \dfrac{3\Pi}{4} \right]^{15} = 1^{15} \text{ cis } \dfrac{45\Pi}{4} = \text{ cis } \dfrac{5\Pi}{4} =$

$-\dfrac{\sqrt{2}}{2} - \dfrac{\sqrt{2}}{2} i$

10) $\left(-\dfrac{\sqrt{3}}{2} - \dfrac{1}{2}i\right)^{50} = \left(1 \text{ cis } \dfrac{7\Pi}{6}\right)^{50} = 1^{50} \text{ cis } \dfrac{175\Pi}{3} = \text{cis } \dfrac{\Pi}{3} = \dfrac{1}{2} + \dfrac{\sqrt{3}}{2}i$

13) $1 + \sqrt{3}\,i = 2 \text{ cis } 60^0;$

$\quad w_k = \sqrt{2} \text{ cis } \left(\dfrac{60^0 + 360^0 k}{2}\right)$ for $k = 0, 1;$

$\quad w_0 = \sqrt{2} \text{ cis } 30^0 = \sqrt{2}\left(\dfrac{\sqrt{3}}{2} + \dfrac{1}{2}i\right) = \dfrac{\sqrt{6}}{2} + \dfrac{\sqrt{2}}{2}i;$

$\quad w_1 = \sqrt{2} \text{ cis } 210^0 = \sqrt{2}\left(-\dfrac{\sqrt{3}}{2} - \dfrac{1}{2}i\right) = -\dfrac{\sqrt{6}}{2} - \dfrac{\sqrt{2}}{2}i$

16) $-8 + 8\sqrt{3}\,i = 16 \text{ cis } 120^0;$

$\quad w_k = \sqrt[4]{16} \text{ cis } \left(\dfrac{120^0 + 360^0 k}{4}\right)$ for $k = 0, 1, 2, 3;$

$\quad w_0 = 2 \text{ cis } 30^0 = 2\left(\dfrac{\sqrt{3}}{2} + \dfrac{1}{2}i\right) = \sqrt{3} + i;$

$\quad w_1 = 2 \text{ cis } 120^0 = 2\left(-\dfrac{1}{2} + \dfrac{\sqrt{3}}{2}i\right) = -1 + \sqrt{3}\,i;$

$\quad w_2 = 2 \text{ cis } 210^0 = 2\left(-\dfrac{\sqrt{3}}{2} - \dfrac{1}{2}i\right) = -\sqrt{3} - i;$

$\quad w_3 = 2 \text{ cis } 300^0 = 2\left(\dfrac{1}{2} - \dfrac{\sqrt{3}}{2}i\right) = 1 - \sqrt{3}\,i$

19) $1 = 1 \text{ cis } 0^0;$

$\quad w_k = \sqrt[6]{1} \text{ cis } \left(\dfrac{0^0 + 360^0 k}{6}\right)$ for $k = 0, 1, ..., 5;$

$\quad w_0 = 1 \text{ cis } 0^0 = 1 + 0i;$

$\quad w_1 = 1 \text{ cis } 60^0 = \dfrac{1}{2} + \dfrac{\sqrt{3}}{2}i;$

$\quad w_2 = 1 \text{ cis } 120^0 = -\dfrac{1}{2} + \dfrac{\sqrt{3}}{2}i;$

$\quad w_3 = 1 \text{ cis } 180^0 = -1 + 0i;$

Exercises 7.4

$w_4 = 1 \text{ cis } 240^0 = -\frac{1}{2} - \frac{\sqrt{3}}{2} i;$

$w_5 = 1 \text{ cis } 300^0 = \frac{1}{2} - \frac{\sqrt{3}}{2} i$

The geometric representation is 6 equispaced points on the unit circle with the first point at $P(1, 0)$.

22) $-\sqrt{3} - i = 2 \text{ cis } 210^0;$

$w_k = \sqrt[5]{2} \text{ cis } \left(\frac{210^0 + 360^0 k}{5} \right)$ for $k = 0, 1, 2, 3, 4;$

$= \sqrt[5]{2} \text{ cis } \theta,$ where $\theta = 42^0, 114^0, 186^0, 258^0, 330^0$

The geometric representation is 5 equispaced points on the unit

circle with the fifth point at $P\left(\frac{\sqrt{3}}{2}, -\frac{1}{2}\right).$

25) $x^6 + 64 = 0 \implies x^6 = -64;$ The problem is now to find the 6 sixth roots of $-64;$ $-64 = -64 + 0i = 64 \text{ cis } 180^0;$

$w_k = \sqrt[6]{64} \text{ cis } \left(\frac{180^0 + 360^0 k}{6} \right)$ for $k = 0, 1, ..., 5;$

$w_0 = 2 \text{ cis } 30^0 = 2 \left[\frac{\sqrt{3}}{2} + \frac{1}{2} i \right] = \sqrt{3} + i;$

$w_1 = 2 \text{ cis } 90^0 = 2(0 + i) = 2i;$

$w_2 = 2 \text{ cis } 150^0 = 2 \left[-\frac{\sqrt{3}}{2} + \frac{1}{2} i \right] = -\sqrt{3} + i;$

$w_3 = 2 \text{ cis } 210^0 = 2 \left[-\frac{\sqrt{3}}{2} - \frac{1}{2} i \right] = -\sqrt{3} - i;$

$w_4 = 2 \text{ cis } 270^0 = 2(0 - i) = -2i;$

$w_5 = 2 \text{ cis } 330^0 = 2 \left[\frac{\sqrt{3}}{2} - \frac{1}{2} i \right] = \sqrt{3} - i$

28) $x^3 - 64i = 0 \implies x^3 = 64i$; The problem is now to find the 3 cube roots of $64i$; $64i = 0 + 64i = 64 \text{ cis } 90^\circ$;

$$w_k = \sqrt[3]{64} \text{ cis } \left[\frac{90^\circ + 360^\circ k}{3}\right] \text{ for } k = 0, 1, 2;$$

$$w_0 = 4 \text{ cis } 30^\circ = 4 \left[\frac{\sqrt{3}}{2} + \frac{1}{2}i\right] = 2\sqrt{3} + 2i;$$

$$w_1 = 4 \text{ cis } 150^\circ = 4 \left[-\frac{\sqrt{3}}{2} + \frac{1}{2}i\right] = -2\sqrt{3} + 2i;$$

$$w_2 = 4 \text{ cis } 270^\circ = 4 (0 - i) = -4i$$

1) $\mathbf{a} + \mathbf{b} = \langle 2, -3 \rangle + \langle 1, 4 \rangle = \langle 3, 1 \rangle$;

$\mathbf{a} - \mathbf{b} = \langle 2, -3 \rangle - \langle 1, 4 \rangle = \langle 1, -7 \rangle$;

$4\mathbf{a} + 5\mathbf{b} = 4 \langle 2, -3 \rangle + 5 \langle 1, 4 \rangle = \langle 8, -12 \rangle + \langle 5, 20 \rangle = \langle 13, 8 \rangle$;

$4\mathbf{a} - 5\mathbf{b} = 4 \langle 2, -3 \rangle - 5 \langle 1, 4 \rangle = \langle 8, -12 \rangle - \langle 5, 20 \rangle = \langle 3, -32 \rangle$

4) Simplifying $\mathbf{a}$ and $\mathbf{b}$ first, $\mathbf{a} = \langle 10, -8 \rangle$ and $\mathbf{b} = \langle -6, 0 \rangle$.

$\mathbf{a} + \mathbf{b} = \langle 10, -8 \rangle + \langle -6, 0 \rangle = \langle 4, -8 \rangle$;

$\mathbf{a} - \mathbf{b} = \langle 10, -8 \rangle - \langle -6, 0 \rangle = \langle 16, -8 \rangle$;

$4\mathbf{a} + 5\mathbf{b} = 4 \langle 10, -8 \rangle + 5 \langle -6, 0 \rangle = \langle 40, -32 \rangle + \langle -30, 0 \rangle$
$= \langle 10, -32 \rangle$;

$4\mathbf{a} - 5\mathbf{b} = 4 \langle 10, -8 \rangle - 5 \langle -6, 0 \rangle = \langle 40, -32 \rangle - \langle -30, 0 \rangle$
$= \langle 70, -32 \rangle$

7) Simplifying $\mathbf{a}$ and $\mathbf{b}$ first, $\mathbf{a} = -4\mathbf{i} + \mathbf{j}$ and $\mathbf{b} = 2\mathbf{i} - 6\mathbf{j}$.

$\mathbf{a} + \mathbf{b} = (-4\mathbf{i} + \mathbf{j}) + (2\mathbf{i} - 6\mathbf{j}) = -2\mathbf{i} - 5\mathbf{j}$;

$\mathbf{a} - \mathbf{b} = (-4\mathbf{i} + \mathbf{j}) - (2\mathbf{i} - 6\mathbf{j}) = -6\mathbf{i} + 7\mathbf{j}$;

$4\mathbf{a} + 5\mathbf{b} = 4(-4\mathbf{i} + \mathbf{j}) + 5(2\mathbf{i} - 6\mathbf{j}) = (-16\mathbf{i} + 4\mathbf{j}) + (10\mathbf{i} - 30\mathbf{j})$
$= -6\mathbf{i} - 26\mathbf{j}$

$4\mathbf{a} - 5\mathbf{b} = 4(-4\mathbf{i} + \mathbf{j}) - 5(2\mathbf{i} - 6\mathbf{j}) = (-16\mathbf{i} + 4\mathbf{j}) - (10\mathbf{i} - 30\mathbf{j})$
$= -26\mathbf{i} + 34\mathbf{j}$

Exercises 7.5

10) Since $a = 0$, $a + b = b = i + j$; $\qquad a - b = -b = -i - j$;

$\qquad\qquad 4a + 5b = 5b = 5i + 5j$; $\qquad 4a - 5b = -5b = -5i - 5j$

13) $a + b = \langle -6, 9 \rangle$; $\quad 2a = \langle -8, 12 \rangle$; $\quad -3b = \langle 6, -9 \rangle$; $\quad$ The

terminal point of each vector is as follows : $\quad a : (-4, 6)$;

$b : (-2, 3)$; $\quad a + b : (-6, 9)$; $\quad 2a : (-8, 12)$; $\quad -3b : (6, -9)$

16) $c - d = \langle 0, 2 \rangle - \langle 0, -1 \rangle = \langle 0, 3 \rangle = \frac{3}{2}\langle 0, 2 \rangle = \frac{3}{2}c$ $\{$ also $-3d\}$

19) $b + d = \langle -1, 0 \rangle + \langle 0, -1 \rangle = \langle -1, -1 \rangle = -\frac{1}{2}e$

22) Let $a = \langle e, f \rangle$.

Proof of the second property :

$(c + d)\,a = (c + d)\,\langle e, f \rangle$

$\qquad = \langle (c + d)\,e, (c + d)\,f \rangle$

$\qquad = \langle ce + de, cf + df \rangle$

$\qquad = \langle ce, cf \rangle + \langle de, df \rangle$

$\qquad = c\langle e, f \rangle + d\langle e, f \rangle$

$\qquad = ca + da$

Proof of the third property :

$(cd)\,a = (cd)\,\langle e, f \rangle$

$\qquad = \langle (cd)\,e, (cd)\,f \rangle$

$\qquad = \langle cde, cdf \rangle$

$\qquad = c\langle de, df \rangle$ $\quad$ or $d\langle ce, cf \rangle$

$\qquad = c\,(d\langle e, f \rangle)$ or $d\,(c\langle e, f \rangle)$

$\qquad = c\,(da)$ $\qquad$ or $d\,(ca)$

Proof of the fourth property :

$1a = 1\,\langle e, f \rangle$

$\qquad = \langle 1e, 1f \rangle$

$\qquad = \langle e, f \rangle$

$\qquad = a$

Proof of the fifth property :

$0a = 0\,\langle e, f \rangle$

$\qquad = \langle 0e, 0f \rangle$

$\qquad = \langle 0, 0 \rangle$

$\qquad = 0$

Also, $c0 = c\,\langle 0, 0 \rangle$

$\qquad = \langle c0, c0 \rangle$

$\qquad = \langle 0, 0 \rangle$

$\qquad = 0$

25) $-(a + b) = -(\langle a_1, a_2 \rangle + \langle b_1, b_2 \rangle) = -(\langle a_1 + b_1, a_2 + b_2 \rangle) =$

$\langle -(a_1 + b_1), -(a_2 + b_2) \rangle = \langle -a_1 - b_1, -a_2 - b_2 \rangle =$

$\langle -a_1, -a_2 \rangle + \langle -b_1, -b_2 \rangle = -a + (-b) = -a - b$

28) Suppose $a + b = a$. Now $a + b = \langle a_1, a_2 \rangle + \langle b_1, b_2 \rangle =$
$\langle a_1 + b_1, a_2 + b_2 \rangle$. So $a_1 + b_1 = a_1$ and $a_2 + b_2 = a_2$, or equivalently,
$b_1 = 0$ and $b_2 = 0$. Now $b = \langle b_1, b_2 \rangle = \langle 0, 0 \rangle = 0$.

31) (a) $|2v| = |2\langle a, b \rangle| = |\langle 2a, 2b \rangle| = \sqrt{(2a)^2 + (2b)^2} = \sqrt{4a^2 + 4b^2}$
$= 2\sqrt{a^2 + b^2} = 2|\langle a, b \rangle| = 2|v|$.

(b) $|\tfrac{1}{2}v| = |\tfrac{1}{2}\langle a, b \rangle| = |\langle \tfrac{1}{2}a, \tfrac{1}{2}b \rangle| = \sqrt{(\tfrac{1}{2}a)^2 + (\tfrac{1}{2}b)^2} = \sqrt{\tfrac{1}{4}a^2 + \tfrac{1}{4}b^2}$
$= \tfrac{1}{2}\sqrt{a^2 + b^2} = \tfrac{1}{2}|\langle a, b \rangle| = \tfrac{1}{2}|v|$.

(c) $|-2v| = |-2\langle a, b \rangle| = |\langle -2a, -2b \rangle| = \sqrt{(-2a)^2 + (-2b)^2} =$
$\sqrt{4a^2 + 4b^2} = 2\sqrt{a^2 + b^2} = 2|\langle a, b \rangle| = 2|v|$.

(d) $|kv| = |k\langle a, b \rangle| = |\langle ka, kb \rangle| = \sqrt{(ka)^2 + (kb)^2} = \sqrt{k^2a^2 + k^2b^2}$
$= \sqrt{k^2}\sqrt{a^2 + b^2} = |k|\,|\langle a, b \rangle| = |k|\,|v|$.

34) $v \cdot v = a_1a_1 + a_2a_2 = (\sqrt{a_1^2 + a_2^2})^2 = |\langle a_1, a_2 \rangle|^2 = |v|^2$

37) $|a| = 3\sqrt{2}$; $\tan \theta = \dfrac{-3}{3} = -1$ and θ in Q IV $\Rightarrow \theta = \dfrac{7\Pi}{4}$

40) $|a| = 10$; $\tan \theta = \tfrac{10}{0}$ (undefined) and the terminal side of θ is on the
positive y-axis $\Rightarrow \theta = \dfrac{\Pi}{2}$

43) $|a| = 18$; $\tan \theta = \tfrac{-18}{0}$ (undefined) and the terminal side of θ is on
the negative y-axis $\Rightarrow \theta = \dfrac{3\Pi}{2}$

46) (a) $= \langle 20 \cos 253^0, 20 \sin 253^0 \rangle \approx \langle -5.85, -19.13 \rangle$;
(b) $= \langle 50 \cos 172^0, 50 \sin 172^0 \rangle \approx \langle -49.51, 6.96 \rangle$;
$a + b \approx \langle -55.36, -12.17 \rangle$; $|a + b| \approx 56.68$ or 57 kg;
$\tan \theta = \tfrac{-12.17}{-55.36} \Rightarrow \theta \approx 192^0$ since θ is in Q III, or equivalently,
S78°W.

49) Let x be the horizontal component and y be the vertical component.

$\cos 35^0 = \frac{x}{50} \Rightarrow x = 50 \cos 35^0 \simeq 40.96;$

$\sin 35^0 = \frac{y}{50} \Rightarrow y = 50 \sin 35^0 \simeq 28.68$

52) Let x be the horizontal component and y be the vertical component.

$\cos 7.5^0 = \frac{x}{160} \Rightarrow x = 160 \cos 7.5^0 \simeq 158.63;$

$\sin 7.5^0 = \frac{y}{160} \Rightarrow y = 160 \sin 7.5^0 \simeq 20.88$

55) (a) $F = F_1 + F_2 = \langle 6 \cos 130^0, 6 \sin 130^0 \rangle +$

$\langle 4 \cos (-120^0), 4 \sin (-120^0) \rangle \simeq \langle -5.86, 1.13 \rangle$

(b) $F + G = 0 \Rightarrow G = -F = \langle 5.86, -1.13 \rangle$

58) (a) Consider the force of 160 pounds to be the resultant vector of two vectors whose initial point is at the astronaut's feet, one along the positive x-axis and the other along the negative y-axis. The angle formed by the resultant vector and the positive x-axis is the complement of θ, $90^0 - \theta$. Now

$\cos (90^0 - \theta) = \frac{\text{x-coordinate}}{160} \Rightarrow \text{x-coordinate} = 160 \sin \theta$ and

$\sin (90^0 - \theta) = \frac{\text{y-coordinate}}{160} \Rightarrow \text{y-coordinate} = 160 \cos \theta.$

{ Remember the cofunction identities : $\cos (90^0 - \theta) = \sin \theta$ and $\sin (90^0 - \theta) = \cos \theta$ }

(b) $27 = 160 \cos \theta \Rightarrow \theta \simeq 80.28^0$ on the moon;

$60 = 160 \cos \theta \Rightarrow \theta \simeq 67.98^0$ on Mars

61) Let **w** represent the wind vector and **p** the plane vector. The direction of the wind is straight north or 90^0 from the positive x-axis. The desired resultant direction is 250^0 or 200^0 from the

positive x-axis. $\mathbf{w} = \langle 50 \cos 90^0, 50 \sin 90^0 \rangle = \langle 0, 50 \rangle$;

$\mathbf{r} = \langle 400 \cos 200^0, 400 \sin 200^0 \rangle \simeq \langle -375.88, -136.81 \rangle$

where $\mathbf{r}$ is the desired resultant of $\mathbf{p} + \mathbf{w}$. Since $\mathbf{r} = \mathbf{p} + \mathbf{w}$,

$\mathbf{p} = \mathbf{r} - \mathbf{w} \simeq \langle -375.88, -186.81 \rangle$; $|\mathbf{p}| \simeq 419.74$ mph;

$\tan \theta \simeq \frac{-186.81}{-375.88}$ and θ is in Q III $\Rightarrow$ $\theta \simeq 206^0$ from the positive

x-axis or 244^0 using the directional form.

64) Let the vectors $\mathbf{c}$, $\mathbf{b}$, and $\mathbf{r}$ denote the current, the boat, and the
resultant, respectively. Let s be the rate of the current and t
the resulting speed. The direction of the boat is N15^{0}E or 75^0 from
the positive x-axis. The current is flowing directly west { 180^0
from the positive x-axis } and the boat is traveling due north { 90^0
from the positive x-axis }.

$\mathbf{b} = \langle 30 \cos 75^0, 30 \sin 75^0 \rangle \simeq \langle 7.76, 28.98 \rangle$;

$\mathbf{c} = \langle s \cos 180^0, s \sin 180^0 \rangle = \langle -s, 0 \rangle$;

$\mathbf{r} = \langle t \cos 90^0, t \sin 90^0 \rangle = \langle 0, t \rangle$;

Since $\mathbf{c} = \mathbf{r} - \mathbf{b}$, we have $-s = 0 - 7.76 \Rightarrow s = 7.76$ or 8 mph.

1) $\gamma = 180^0 - \alpha - \beta = 75^0$;

$\dfrac{a}{\sin \alpha} = \dfrac{b}{\sin \beta} \Rightarrow a = \dfrac{b \sin \alpha}{\sin \beta} = \dfrac{100 \cdot \frac{\sqrt{3}}{2}}{\sqrt{2}/2} = 50\sqrt{6}$;

$\dfrac{c}{\sin \gamma} = \dfrac{b}{\sin \beta} \Rightarrow c = \dfrac{b \sin \gamma}{\sin \beta} = \dfrac{100 \sin (45^0 + 30^0)}{\sqrt{2}/2} =$

{ Use the identity $\sin (\alpha + \beta) = \sin \alpha \cos \beta + \cos \alpha \sin \beta$ }

$\dfrac{200}{\sqrt{2}} \left(\dfrac{\sqrt{2}}{2} \cdot \dfrac{\sqrt{3}}{2} + \dfrac{\sqrt{2}}{2} \cdot \dfrac{1}{2} \right) = 50 \ (1 + \sqrt{3})$

4) $\alpha = \cos^{-1}\left(\dfrac{b^2 + c^2 - a^2}{2bc}\right) = \cos^{-1}\left(\dfrac{7}{8}\right)$;

$\beta = \cos^{-1}\left(\dfrac{a^2 + c^2 - b^2}{2ac}\right) = \cos^{-1}\left(\dfrac{11}{16}\right)$;

$\gamma = \cos^{-1}\left(\dfrac{a^2 + b^2 - c^2}{2ab}\right) = \cos^{-1}\left(-\dfrac{1}{4}\right)$

7) $b = \sqrt{a^2 + c^2 - 2ac\cos\beta} \approx \sqrt{102.8} \approx 10.1$;

$\alpha = \cos^{-1}\left(\dfrac{b^2 + c^2 - a^2}{2bc}\right) \approx \cos^{-1}(0.9116) \approx 24^0$;

$\gamma = 180^0 - \alpha - \beta \approx 41^0$

10) Using Heron's Formula with a = 4, b = 7, and c = 10, we have

$\quad s = \frac{1}{2}(a + b + c) = \frac{1}{2}(4 + 7 + 10) = 10.5$;

$\quad A = \sqrt{s(s - a)(s - b)(s - c)} = \sqrt{(10.5)(6.5)(3.5)(0.5)} \approx$

$\quad\quad 10.9$ square units

For Exercises 13-22, $\cos\theta + i\sin\theta$ will be represented by cis θ.

13) $r = 17$; $\tan\theta = \frac{0}{17} = 0 \Rightarrow \theta = \Pi$; 17 cis Π

16) $r = \sqrt{41}$; $\tan\theta = \frac{5}{4}$ and θ in Q I $\Rightarrow \theta = \tan^{-1}\frac{5}{4}$;

$\quad \sqrt{41}$ cis $(\tan^{-1}\frac{5}{4})$

19) $(3 - 3i)^5 = \left(3\sqrt{2} \text{ cis } \dfrac{7\Pi}{4}\right)^5 = (3\sqrt{2})^5 \text{ cis } \dfrac{35\Pi}{4} = 972\sqrt{2} \text{ cis } \dfrac{3\Pi}{4}$

$\quad = 972\sqrt{2}\left(-\dfrac{\sqrt{2}}{2} + \dfrac{\sqrt{2}}{2}i\right) = -972 + 972i$

22) $x^5 - 32 = 0 \Rightarrow x^5 = 32$; The problem is now to find the 5 fifth

$\quad$ roots of 32; $32 = 32 + 0i = 32$ cis 0^0;

$\quad w_k = \sqrt[5]{32} \text{ cis }\left(\dfrac{0^0 + 360^0 k}{5}\right)$ for k = 0, 1, 2, 3, 4;

$\quad w_k = 2$ cis θ, where $\theta = 0^0, 72^0, 144^0, 216^0, 288^0$

25) $|r - a| = c \iff |\langle x - a_1, y - a_2\rangle| = c \iff$

$\underline{\sqrt{(x - a_1)^2 + (y - a_2)^2}} = c \iff (x - a_1)^2 + (y - a_2)^2 = c^2;$

This is a circle with center (a_1, a_2) and radius c.

28) $S60°E$ is equivalent to $330°$ and $N74°E$ is equivalent to $16°$.

{ where both $330°$ and $16°$ are measured from the positive x-axis }

$\langle 72 \cos 330°, 72 \sin 330°\rangle + \langle 46 \cos 16°, 46 \sin 16°\rangle = r \simeq$

$\langle 106.57, -23.32\rangle;$ $|r| \simeq 109\,kg;$ $\tan\theta \simeq \frac{-23.32}{106.57} \implies \theta \simeq -12°$ or

equivalently, $S78°E$.

31) Let P be the point at the base of the shorter building, S the point at the top of the shorter building, T the point at the top of the taller building, and Q the point 50 feet up the side of the taller building.

(a) $\angle SPT = 90° - 62° = 28°;$ $\angle PST = 90° + 59° = 149°;$ Thus

$\angle STP = 180° - 28° - 149° = 3°;$

$\dfrac{\overline{ST}}{\sin 28°} = \dfrac{50}{\sin 3°} \implies \overline{ST} \simeq 448.5 \text{ feet}$

(b) $h = \overline{QT} + 50 = \overline{ST} \sin 59° + 50 \simeq 434.45 \text{ feet}$

34) Let d be the distance each girl walks before losing contact with each other. $d = 5t$ where t is in hours. Using the Law of Cosines,

$10^2 = d^2 + d^2 - 2\,(d)\,(d)\cos 105° \implies 100 = (2 - 2\cos 105°)\,d^2 \implies$

$d \simeq 6.30 \implies t \simeq 1.26$ hours or 1 hour and 16 minutes.

Exercises 8.1

The notation E1 and E2 refers to equation 1 and equation 2.

1) Substituting y in E2 into E1 yields $2x - 1 = x^2 - 4 \Rightarrow$
$x^2 - 2x - 3 = 0 \Rightarrow (x - 3)(x + 1) = 0 \Rightarrow x = 3, -1; \ y = 5, -3;$
$(3, 5)$ and $(-1, -3)$

4) Solving E2 for x and substituting into E1 yields $y^2 = -2y - 3 \Rightarrow$
$y^2 + 2y + 3 = 0 \Rightarrow y = \dfrac{-2 \pm \sqrt{4 - 12}}{2}$; There are no real solutions
since the discriminant is negative.

7) Solving E1 for x and substituting into E2 yields $2(-2y - 1) - 3y =$
$12 \Rightarrow -7y = 14 \Rightarrow y = -2; \ x = 3; \ (3, -2)$

10) Solving E1 for x and substituting into E2 yields
$8(\frac{1}{2} + \frac{5}{4}y) - 10y = -5 \Rightarrow 4 = -5;$ This is a contradiction,
therefore, there are no solutions.

13) Solving E2 for y and substituting into E1 yields $x^2 + (x + 4)^2 = 8$
$\Rightarrow 2x^2 + 8x + 8 = 0 \Rightarrow 2(x + 2)^2 = 0 \Rightarrow x = -2; \ y = 2;$
$(-2, 2)$

16) Solving E2 for y and substituting into E1 yields
$x^2 + (-2x - 1)^2 = 16 \Rightarrow 5x^2 + 4x - 15 = 0 \Rightarrow$
$x = \dfrac{-4 \pm \sqrt{16 + 300}}{10} = \dfrac{-2 \pm \sqrt{79}}{5};$
$y = -2\left(\dfrac{-2 \pm \sqrt{79}}{5}\right) - 1 = \dfrac{4 \mp 2\sqrt{79}}{5} - \dfrac{5}{5} = \dfrac{-1 \mp 2\sqrt{79}}{5};$
$\left(\dfrac{-2 + \sqrt{79}}{5}, \dfrac{-1 - 2\sqrt{79}}{5}\right)$ and $\left(\dfrac{-2 - \sqrt{79}}{5}, \dfrac{-1 + 2\sqrt{79}}{5}\right)$

19) Solving E2 for x and substituting into E1 yields
$(1 - y - 1)^2 + (y + 2)^2 = 10 \Rightarrow 2y^2 + 4y - 6 = 0 \Rightarrow$
$2(y + 3)(y - 1) = 0 \Rightarrow y = -3, 1; \ x = 4, 0; \ (4, -3)$ and $(0, 1)$

22) Solving E2 for x and substituting into E1 yields

$y + 1 = y^2 - 4y + 5 \Rightarrow y^2 - 5y + 4 = 0 \Rightarrow (y - 1)(y - 4) = 0$

$\Rightarrow y = 1, 4; \; x = 2, 5; \; \underline{(2, 1) \text{ and } (5, 4)}$

25) Solving E1 for x^2 and substituting into E2 yields $(y^2 + 4) + y^2 = 12$

$\Rightarrow 2y^2 = 8 \Rightarrow y = \pm 2; \; x = \pm 2\sqrt{2}; \; \underline{(\pm 2\sqrt{2}, \pm 2) \text{ Four solutions.}}$

28) Solving E2 for z^2 and substituting into E1 yields

$2x - 3y - (x - y + 1) = 0 \Rightarrow \begin{cases} x - 2y = 1 & (E3) \\ x^2 - xy = 0 & (E4) \end{cases}$

Now solve E3 for x and substitute into E4 yielding

$(2y + 1)^2 - (2y + 1)y = 0 \Rightarrow 2y^2 + 3y + 1 = 0 \Rightarrow$

$(2y + 1)(y + 1) = 0 \Rightarrow y = -\tfrac{1}{2} \text{ or } -1; \; x = 2y + 1 = 0 \text{ or } -1;$

$z^2 = x - y + 1 = \tfrac{3}{2} \text{ or } 1 \Rightarrow z = \pm\sqrt{3/2} \text{ or } \pm 1;$

$(0, -\tfrac{1}{2}, \pm\sqrt{3/2}) \text{ and } (-1, -1, \pm 1) \; \underline{\text{Four solutions.}}$

31) Using $P = 2\ell + 2w$ and $A = \ell w$, we have

$\begin{cases} 2\ell + 2w = 40 & (E1) \\ \ell w \quad\;\;\; = 96 & (E2) \end{cases}$

Solving E1 for ℓ and substituting into E2 yields $(20 - w)w = 96$

$\Rightarrow w^2 - 20w + 96 = 0 \Rightarrow (w - 8)(w - 12) = 0 \Rightarrow w = 8, 12;$

$\ell = 12, 8;$ The rectangle is $\underline{12 \text{ in. by } 8 \text{ in.}}$

34) From the graph it is clear that
there are no intersection points
of the two graphs, and therefore
there are no solutions.

See Figure 34.

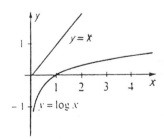

Figure 34

Exercises 8.1

37) (a) Let S = 40,000 and R = 60,000 for the data for 1983 and 1984. Then let S = 60,000 and R = 72,000 for the data for 1984 and 1985.

$$\begin{cases} 60,000 = \dfrac{40,000\,a}{40,000 + b} \\[2mm] 72,000 = \dfrac{60,000\,a}{60,000 + b} \end{cases} \Rightarrow \begin{cases} 120,000 + 3b = 2a \quad (E1) \\ 360,000 + 6b = 5a \quad (E2) \end{cases}$$

Solving E1 for a and substituting into E2 yields $360,000 + 6b = 5\,(60,000 + \tfrac{3}{2}b) \Rightarrow 60,000 = \tfrac{3}{2}b \Rightarrow b = 40,000$; a = 120,000

(b) Now let S = 72,000 so R = $\dfrac{(120,000)\,(72,000)}{72,000 + 40,000} = \dfrac{540,000}{7} \approx 77,143$

40) Assume that the fencing will be constructed in a rectangular shape. Let ℓ be the length of the side opposite the river and w the length of the other two sides. (10 acres = 435,600 ft²)

$$\begin{cases} \ell + 2w = 2,420 \quad (E1) \\ \ell w = 435,600 \quad (E2) \end{cases}$$

Solve E1 for ℓ and substitute into E2.

$(2,420 - 2w)\,w = 435,600 \Rightarrow 2w^2 - 2,420w + 435,600 = 0 \Rightarrow$
$w^2 - 1,210w + 217,800 = 0 \Rightarrow (w - 990)\,(w - 220) = 0 \Rightarrow$
$w = 990, 220;\ \ \ell = 2,420 - 2w = 440, 1980;$ Either a 990 ft by 440 ft or a 220 ft by 1980 ft rectangle will enclose 10 acres.

Exercises 8.2

The notation E1 and E2 refers to equation 1 and equation 2. 3E2 symbolizes "3 times equation 2". After the value of one variable is found, the value(s) of the other variable(s) will be stated and can be found by substituting the known value(s) back into the original equations.

1) $-2\,E2 + E1 \Rightarrow 7y = -14 \Rightarrow y = -2;\ x = 4$

4) $4\,E1 - 7\,E2 \Rightarrow -53y = 106 \Rightarrow y = -2;\ x = -1$

7) $3\,E1 - 5\,E2 \Rightarrow -53y = -28 \Rightarrow y = \frac{28}{53};$

Instead of substituting into one of the equations to find the value of the other variable, it is usually easier to pick different multipliers and resolve the system for the other variable. { This is especially true when the substitution would involve working with fractions. }

$7\,E1 + 6\,E2 \Rightarrow 53x = 76 \Rightarrow x = \frac{76}{53}$

10) When equations contain fractions, it is usually easier to solve a system obtained by multiplying each equation by its least common denominator.

$$\left.\begin{array}{r} 10\,E1 \\ 12\,E2 \end{array}\right\} \Rightarrow \left\{\begin{array}{ll} 5t - 2v = 15 & (E3) \\ 8t + 3v = 5 & (E4) \end{array}\right.$$

$8\,E3 - 5\,E4 \Rightarrow -31v = 95 \Rightarrow v = -\frac{95}{31};$

$3\,E3 + 2\,E4 \Rightarrow 31t = 55 \Rightarrow t = \frac{55}{31}$

13) $3\,E1 + E2 \Rightarrow 0 = 27;$ Contradiction, $\therefore$ <u>no solution.</u>

16) $-3\,E1 + E2 \Rightarrow 0 = 0;$ This is an identity. The solution is all ordered pairs (x, y) such that $x - 5y = 2$.

19) Using the hint, the system is $\left\{\begin{array}{ll} 2x' + 3y' = -2 & (E3) \\ 4x' - 5y' = 1 & (E4) \end{array}\right.$

$-2\,E3 + E4 \Rightarrow -11y' = 5 \Rightarrow y' = -\frac{5}{11};$

$5\,E3 + 3\,E4 \Rightarrow 22x' = -7 \Rightarrow x' = -\frac{7}{22};$

Resubstituting, we have $x = -\frac{22}{7}$ and $y = -\frac{11}{5}$.

Exercises 8.2

22) Let x be the number of passengers that purchased a ticket to Phoenix and y the number of passengers that purchased a ticket to Albuquerque.

$\underline{\text{quantity}} \begin{cases} x + y = 185 \\ 45x + 60y = 10{,}500 \end{cases}$
$\underline{\text{value}}$

E2 − 45 E1 $\Rightarrow$ 15y = 2,175 $\Rightarrow$ y = 145; x = 40

25) $\underline{\text{perimeter}} \begin{cases} 2y + 2\Pi \left(\frac{1}{2} x\right) = 40 \\ xy = 2 \left(\Pi \left(\frac{1}{2} x\right)^2\right) \end{cases}$
$\underline{\text{area}}$

Solving E1 for y and substituting into E2 yields

$$x \left(\frac{40 - \Pi x}{2}\right) = \frac{\Pi x^2}{2} \Rightarrow 40x = 2\Pi x^2 \Rightarrow 2x \left(\Pi x - 20\right) = 0 \Rightarrow$$

$x = 0, \frac{20}{\Pi}; \quad x = \frac{20}{\Pi}$ ft. and y = 10 ft.

28) Let x be the flow rate of the inlet pipe and y the flow rate of one of the outlet pipes. (inlet rate − outlet rate) (hours) = gallons

$\underline{\text{both open}} \begin{cases} (x - 2y)\, 5 = 300 \\ (x - y)\, 3 = 300 \end{cases} \Rightarrow \begin{cases} x - 2y = 60 \\ x - y = 100 \end{cases}$
$\underline{\text{one closed}}$

E1 − E2 $\Rightarrow$ −y = −40 $\Rightarrow$ y = 40 gal/hr; x = 140 gal/hr

31) Let x be the speed of the plane and y the speed of the wind.

$$\text{distance} = \text{(rate) (time)}$$

$\underline{\text{with the wind}} \begin{cases} 1200 = (x + y)\,(2) \\ 1200 = (x - y)\,(2.5) \end{cases} \Rightarrow \begin{cases} 600 = x + y \\ 480 = x - y \end{cases}$
$\underline{\text{against the wind}}$

E1 + E2 $\Rightarrow$ 2x = 1,080 $\Rightarrow$ x = 540 mph; y = 60 mph

34) s (1) = 84 $\Rightarrow$ 84 = −16 + v_0 + s_0; (E1)

s (2) = 116 $\Rightarrow$ 116 = −64 + 2v_0 + s_0; (E2)

E2 − E1 $\Rightarrow$ v_0 = 80; s_0 = 20; s (t) = −16t² + 80t + 20

37) (a) $6x + 5y$ represents the total bill for the plumber's business.

This should equal the plumber's income which is $(6 + 4)x$.

$4x + 6y$ represents the total bill for the electrician's business.

This should equal the electrician's income which is $(5 + 6)y$.

$$\begin{cases} 6x + 5y = 10x \\ 4x + 6y = 11y \end{cases} \Rightarrow \begin{cases} 5y = 4x \\ 4x = 5y \end{cases} \Rightarrow y = \tfrac{4}{5}x \text{ or } 0.80x$$

(b) The electrician should charge 80% of what the plumber charges.

80% of $20 per hour is $16 per hour.

1) $\begin{bmatrix} 1 & -2 & -3 & -1 \\ 2 & 1 & 1 & 6 \\ 1 & 3 & -2 & 13 \end{bmatrix} \begin{array}{l} R_2 - 2R_1 \\ R_3 - R_1 \end{array} \longrightarrow \begin{bmatrix} 1 & -2 & -3 & -1 \\ 0 & 5 & 7 & 8 \\ 0 & 5 & 1 & 14 \end{bmatrix} R_3 - R_2 \longrightarrow$

$\begin{bmatrix} 1 & -2 & -3 & -1 \\ 0 & 5 & 7 & 8 \\ 0 & 0 & -6 & 6 \end{bmatrix} -\tfrac{1}{6}R_3 \Rightarrow z = -1$

$R_2 : 5y + 7z = 8 \Rightarrow 5y - 7 = 8 \Rightarrow y = 3$

$R_1 : x - 2y - 3z = -1 \Rightarrow x - 6 + 3 = -1 \Rightarrow x = 2$

4) $\begin{bmatrix} 4 & -1 & 3 & 6 \\ -8 & 3 & -5 & -6 \\ 5 & -4 & 0 & -9 \end{bmatrix} R_3 - R_1 \longrightarrow \begin{bmatrix} 4 & -1 & 3 & 6 \\ -8 & 3 & -5 & -6 \\ 1 & -3 & -3 & -15 \end{bmatrix} \begin{array}{l} R_1 - 4R_3 \\ R_2 + 8R_3 \end{array} \longrightarrow$

$\begin{bmatrix} 0 & 11 & 15 & 66 \\ 0 & -21 & -29 & -126 \\ 1 & -3 & -3 & -15 \end{bmatrix} R_2 + 2R_1 \longrightarrow \begin{bmatrix} 0 & 11 & 15 & 66 \\ 0 & 1 & 1 & 6 \\ 1 & -3 & -3 & -15 \end{bmatrix} R_1 - 11R_2 \longrightarrow$

(continued on next page)

$$\begin{bmatrix} 0 & 0 & 4 & 0 \\ 0 & 1 & 1 & 6 \\ 1 & -3 & -3 & -15 \end{bmatrix} \tfrac{1}{4}R_1 \implies z = 0$$

$R_2 : y + z = 6 \implies y = 6$

$R_3 : x - 3y - 3z = -15 \implies x = 3$

7) $\begin{bmatrix} -3 & 2 & 1 & 1 \\ 4 & 1 & -3 & 4 \\ 2 & -3 & 2 & -3 \end{bmatrix} \xrightarrow{R_1 + R_2} \begin{bmatrix} 1 & 3 & -2 & 5 \\ 4 & 1 & -3 & 4 \\ 2 & -3 & 2 & -3 \end{bmatrix} \begin{matrix} R_2 - 4R_1 \\ R_3 - 2R_1 \end{matrix} \xrightarrow{}$

$\begin{bmatrix} 1 & 3 & -2 & 5 \\ 0 & -11 & 5 & -16 \\ 0 & -9 & 6 & -13 \end{bmatrix} \xrightarrow{4R_2 - 5R_3} \begin{bmatrix} 1 & 3 & -2 & 5 \\ 0 & 1 & -10 & 1 \\ 0 & -9 & 6 & -13 \end{bmatrix} \xrightarrow{R_3 + 9R_2}$

$\begin{bmatrix} 1 & 3 & -2 & 5 \\ 0 & 1 & -10 & 1 \\ 0 & 0 & -84 & -4 \end{bmatrix} -\tfrac{1}{84}R_3 \implies z = \tfrac{1}{21}$

$R_2 : y - 10z = 1 \implies y = \tfrac{31}{21}$

$R_1 : x + 3y - 2z = 5 \implies x = \tfrac{14}{21} = \tfrac{2}{3}$

10) $\begin{bmatrix} 1 & -1 & -2 & 0 \\ 2 & -1 & 1 & 0 \\ 2 & -3 & -1 & 0 \end{bmatrix} \begin{matrix} R_2 - 2R_1 \\ R_3 - 2R_1 \end{matrix} \xrightarrow{} \begin{bmatrix} 1 & -1 & -2 & 0 \\ 0 & 1 & 5 & 0 \\ 0 & -1 & 3 & 0 \end{bmatrix} \xrightarrow{R_3 + R_2}$

$\begin{bmatrix} 1 & -1 & -2 & 0 \\ 0 & 1 & 5 & 0 \\ 0 & 0 & 8 & 0 \end{bmatrix} \tfrac{1}{8}R_3 \implies z = 0$

$R_2 : y + 5z = 0 \implies y = 0; \qquad R_1 : x - y - 2z = 0 \implies x = 0$

13) $\begin{bmatrix} 1 & 4 & -1 & -2 \\ 3 & -2 & 5 & 7 \end{bmatrix} \xrightarrow{R_2 - 3R_1} \begin{bmatrix} 1 & 4 & -1 & -2 \\ 0 & -14 & 8 & 13 \end{bmatrix}$

$R_2 : -14y + 8z = 13 \implies y = \frac{4}{7}z - \frac{13}{14}$

$R_1 : x + 4y - z = -2 \implies x = -4\left(\frac{4}{7}z - \frac{13}{14}\right) + z - 2 = -\frac{9}{7}z + \frac{12}{7}$

The solution is $\left(-\frac{9}{7}c + \frac{12}{7}, \frac{4}{7}c - \frac{13}{14}, c\right)$ where c is any real number.

16) $\begin{bmatrix} 5 & 2 & -1 & 10 \\ 0 & 1 & 1 & -3 \end{bmatrix}$

$R_2 : y + z = -3 \implies y = -z - 3$

$R_1 : 5x + 2y - z = 10 \implies x = -\frac{2}{5}(-z - 3) + \frac{1}{5}z + 2 = \frac{3}{5}z + \frac{16}{5}$

The solution is $\left(\frac{3}{5}c + \frac{16}{5}, -c - 3, c\right)$ where c is any real number.

19) $\begin{bmatrix} 1 & 3 & -2 & 1 & -2 & -5 \\ 2 & -1 & -2 & 2 & -5 & 2 \\ -1 & 4 & 2 & -3 & 8 & -4 \\ 3 & -2 & -4 & 1 & -3 & -3 \\ 4 & -6 & 1 & -2 & 1 & 10 \end{bmatrix} \begin{matrix} \\ R_2 - 2R_1 \\ R_3 + R_1 \\ R_4 - 3R_1 \\ R_5 - 4R_1 \end{matrix}$

$\begin{bmatrix} 1 & 3 & -2 & 1 & -2 & -5 \\ 0 & -7 & 2 & 0 & -1 & 12 \\ 0 & 7 & 0 & -2 & 6 & -9 \\ 0 & -11 & 2 & -2 & 3 & 12 \\ 0 & -18 & 9 & -6 & 9 & 30 \end{bmatrix} \begin{matrix} \\ \\ 3R_2 - 2R_4 \\ \\ -\frac{1}{3}R_5 \end{matrix}$

(continued on next page)

Exercises 8.3

$$\begin{bmatrix} 1 & 3 & -2 & 1 & -2 & -5 \\ 0 & 1 & 2 & 4 & -9 & 12 \\ 0 & 7 & 0 & -2 & 6 & -9 \\ 0 & -11 & 2 & -2 & 3 & 12 \\ 0 & 6 & -3 & 2 & -3 & -10 \end{bmatrix} \begin{array}{l} \\ \\ R_3 - 7R_2 \\ R_4 + 11R_2 \\ R_5 - 6R_2 \end{array} \longrightarrow$$

$$\begin{bmatrix} 1 & 3 & -2 & 1 & -2 & -5 \\ 0 & 1 & 2 & 4 & -9 & 12 \\ 0 & 0 & -14 & -30 & 69 & -93 \\ 0 & 0 & 24 & 42 & -96 & 144 \\ 0 & 0 & -15 & -22 & 51 & -82 \end{bmatrix} \begin{array}{l} \\ \\ R_3 - R_5 \\ \frac{1}{2}R_4 \end{array} \longrightarrow$$

$$\begin{bmatrix} 1 & 3 & -2 & 1 & -2 & -5 \\ 0 & 1 & 2 & 4 & -9 & 12 \\ 0 & 0 & 1 & -8 & 18 & -11 \\ 0 & 0 & 12 & 21 & -48 & 72 \\ 0 & 0 & -15 & -22 & 51 & -82 \end{bmatrix} \begin{array}{l} \\ \\ \\ R_4 - 12R_3 \\ R_5 + 15R_3 \end{array} \longrightarrow$$

$$\begin{bmatrix} 1 & 3 & -2 & 1 & -2 & -5 \\ 0 & 1 & 2 & 4 & -9 & 12 \\ 0 & 0 & 1 & -8 & 18 & -11 \\ 0 & 0 & 0 & 117 & -264 & 204 \\ 0 & 0 & 0 & -142 & 321 & -247 \end{bmatrix} \begin{array}{l} \\ \\ \\ \\ 17R_4 + 14R_5 \end{array} \longrightarrow$$

(continued on next page)

$$\begin{bmatrix} 1 & 3 & -2 & 1 & -2 & -5 \\ 0 & 1 & 2 & 4 & -9 & 12 \\ 0 & 0 & 1 & -8 & 18 & -11 \\ 0 & 0 & 0 & 1 & 6 & 10 \\ 0 & 0 & 0 & -142 & 321 & -247 \end{bmatrix} \xrightarrow{R_5 + 142\,R_4}$$

$$\begin{bmatrix} 1 & 3 & -2 & 1 & -2 & -5 \\ 0 & 1 & 2 & 4 & -9 & 12 \\ 0 & 0 & 1 & -8 & 18 & -11 \\ 0 & 0 & 0 & 1 & 6 & 10 \\ 0 & 0 & 0 & 0 & 1173 & 1173 \end{bmatrix} \xrightarrow{\frac{1}{1173} R_5} \Rightarrow t = 1$$

$R_4 : s + 6t = 10 \Rightarrow s = 4$

$R_3 : z - 8s + 18t = -11 \Rightarrow z = 3$

$R_2 : y + 2z + 4s - 9t = 12 \Rightarrow y = -1$

$R_1 : x + 3y - 2z + s - 2t = -5 \Rightarrow x = 2$

22) $\begin{bmatrix} 2 & -5 & 0 & 4 \\ 0 & 3 & 2 & -3 \\ 7 & 0 & -3 & 1 \end{bmatrix} \xrightarrow{4\,R_1 - R_3} \begin{bmatrix} 1 & -20 & 3 & 15 \\ 0 & 3 & 2 & -3 \\ 7 & 0 & -3 & 1 \end{bmatrix} \xrightarrow{R_3 - 7\,R_1}$

$$\begin{bmatrix} 1 & -20 & 3 & 15 \\ 0 & 3 & 2 & -3 \\ 0 & 140 & -24 & -104 \end{bmatrix} \xrightarrow{47\,R_2 - R_3}$$

$$\begin{bmatrix} 1 & -20 & 3 & 15 \\ 0 & 1 & 118 & -37 \\ 0 & 140 & -24 & -104 \end{bmatrix} \xrightarrow{\frac{1}{4}R_3 - 35\,R_2} \qquad \text{(continued on next page)}$$

$$\begin{bmatrix} 1 & -20 & 3 & 15 \\ 0 & 1 & 118 & -37 \\ 0 & 0 & -4136 & 1269 \end{bmatrix} - \frac{1}{4136} R_3 \implies z = -\frac{1269}{4136} = -\frac{27}{88}$$

$R_2 : y + 118z = -37 \implies y = -\frac{70}{88}$

$R_1 : x - 20y + 3z = 15 \implies x = \frac{1}{88}$

25) $\begin{bmatrix} 1 & -3 & 4 \\ 2 & 3 & 5 \\ 1 & 1 & -2 \end{bmatrix} \begin{array}{l} R_2 - 2R_1 \\ \\ R_3 - R_1 \end{array} \begin{bmatrix} 1 & -3 & 4 \\ 0 & 9 & -3 \\ 0 & 4 & -6 \end{bmatrix} R_2 - 2R_3$

$\begin{bmatrix} 1 & -3 & 4 \\ 0 & 1 & 9 \\ 0 & 4 & -6 \end{bmatrix} \begin{array}{l} R_1 + 3R_2 \\ \\ R_3 - 4R_2 \end{array} \begin{bmatrix} 1 & 0 & 31 \\ 0 & 1 & 9 \\ 0 & 0 & -42 \end{bmatrix}$

There is a contradiction in row 3, therefore there is <u>no solution</u>.

28) Let z be the number of hours it takes for pipe C to fill the pool alone. By adding the hourly rates { i.e. how much each pipe fills in one hour }, we have $\frac{1}{8} + \frac{1}{z} = \frac{1}{6} \implies z = 24$; Let y be the number of hours it takes for pipe B to fill the pool alone. $\frac{1}{y} + \frac{1}{24} = \frac{1}{10} \implies$ $y = \frac{120}{7}$; Let x be the number of hours it takes for all three pipes to fill the pool. $\frac{1}{8} + \frac{7}{120} + \frac{1}{24} = \frac{1}{x} \implies x = \frac{120}{27} = \frac{40}{9}$

31) Let x be the amount of G_1, y the amount of G_2, and z the amount of G_3.

<u>quantity</u> $x + y + z = 600$ (E1)

<u>quality</u> $0.30x + 0.20y + 0.15z = (0.25)(600)$ (E2)

<u>constraint</u> $z = 100 + y$ (E3)

Substitute $z = 100 + y$ into E1 and 100 E2 to obtain

$$\begin{cases} x + 2y = 500 & \text{(E4)} \\ 30x + 35y = 13{,}500 & \text{(E5)} \end{cases}$$

E5 $-$ 30 E4 $\Rightarrow$ $-25y = -1{,}500$ $\Rightarrow$ $y = 60$; $x = 380$; $z = 160$

34) Let x be the number of birds on island A, y the number of birds on island B, and z the number of birds on island C.

$$\begin{array}{ll} \underline{\text{quantity}} & \left\{\begin{array}{l} x + \qquad y + \qquad z = 35{,}000 \quad \text{(E1)} \\ \underline{\text{island A}} \quad x - 0.10x + 0.05z = x \qquad\qquad \text{(E2)} \\ \underline{\text{island B}} \quad y - 0.20y + 0.10x = y \qquad\qquad \text{(E3)} \\ \underline{\text{island C}} \quad z - 0.05z + 0.20y = z \qquad\qquad \text{(E4)} \end{array}\right. \end{array}$$

Solve E2 for z ($z = 2x$), E3 for y ($y = \frac{1}{2}x$), and substitute both expressions into E1 to obtain $x + \frac{1}{2}x + 2x = 35{,}000 \Rightarrow$
$\frac{7}{2}x = 35{,}000 \Rightarrow x = 10{,}000$; $y = 5{,}000$; $z = 20{,}000$

37) The general equation is $f(x) = ax^2 + bx + c$.

$$\begin{cases} f(1) = 3 \\ f(2) = \frac{5}{2} \\ f(-1) = 1 \end{cases} \Rightarrow \begin{cases} 3 = a + b + c & \text{(E1)} \\ \frac{5}{2} = 4a + 2b + c & \text{(E2)} \\ 1 = a - b + c & \text{(E3)} \end{cases}$$

Solving E3 for c and substituting into E1 and E2 yields

$$\begin{cases} 3 = a + b + (1 - a + b) \\ \frac{5}{2} = 4a + 2b + (1 - a + b) \end{cases} \Rightarrow \begin{cases} 2 = 2b \\ \frac{3}{2} = 3a + 3b \end{cases}$$

Since $b = 1$, we obtain $a = -\frac{1}{2}$ and $c = \frac{5}{2}$

40) $\begin{array}{l} \underline{P_1} \\ \underline{P_2} \\ \underline{P_3} \end{array}$ $\begin{cases} 9a + 3b + c = -1 & \text{(E1)} \\ a + b + c = -7 & \text{(E2)} \\ 4a - 2b + c = 14 & \text{(E3)} \end{cases}$

(continued on next page)

Exercises 8.3

Solving E2 for c and substituting into E1 and E3 yields

$$\begin{cases} 9a + 3b - 7 - a - b = -1 \\ 4a - 2b - 7 - a - b = 14 \end{cases} \Rightarrow \begin{cases} 4a + b = 3 \quad \text{(E4)} \\ a - b = 7 \quad \text{(E5)} \end{cases}$$

$$\text{E4} + \text{E5} \Rightarrow 5a = 10 \Rightarrow a = 2; \ b = -5; \ c = -4$$

Exercises 8.4

The general outline for the solutions in this section is as follows :

1st line) The expression is shown on the left side of the equation and
its decomposition is on the right side.

2nd line) The equation in the first line is multiplied by its least common
denominator and left in factored form.

3rd line and beyond) Values are substituted into the equation in the
second line and the coefficients are solved for.
It will be stated when the method of equating
coefficients is used.

1) $\dfrac{8x - 1}{(x - 2)\,(x + 3)} = \dfrac{A}{x - 2} + \dfrac{B}{x + 3}$

$8x - 1 = A\,(x + 3) + B\,(x - 2)$

$x = -3 : -25 = -5B \Rightarrow B = 5$

$x = \ \ 2 : \ \ 15 = 5A \Rightarrow A = 3$

4) $\dfrac{5x - 12}{x\,(x - 4)} = \dfrac{A}{x} + \dfrac{B}{x - 4}$

$5x - 12 : A\,(x - 4) + Bx$

$x = 4 : \ \ \ 8 = 4B \Rightarrow B = 2$

$x = 0 : -12 = -4A \Rightarrow A = 3$

7) $\dfrac{4x^2 - 5x - 15}{x\,(x - 5)\,(x + 1)} = \dfrac{A}{x} + \dfrac{B}{x - 5} + \dfrac{C}{x + 1}$

$4x^2 - 5x - 15 = A\,(x - 5)\,(x + 1) + Bx\,(x + 1) + Cx\,(x - 5)$

$x = -1 : \quad -6 = \quad 6C \Rightarrow C = -1$

$x = \quad 0 : -15 = -5A \Rightarrow A = \quad 3$

$x = \quad 5 : \quad 60 = 30B \Rightarrow B = \quad 2$

10) $\dfrac{5x^2 - 4}{x^2\,(x + 2)} = \dfrac{A}{x} + \dfrac{B}{x^2} + \dfrac{C}{x + 2}$

$5x^2 - 4 = Ax\,(x + 2) + B\,(x + 2) + Cx^2$

$x = -2 : 16 = 4C \qquad\qquad \Rightarrow C = \quad 4$

$x = \quad 0 : -4 = 2B \qquad\qquad \Rightarrow B = -2$

$x = \quad 1 : \quad 1 = 3A + 3B + C \Rightarrow A = \quad 1$

13) $\dfrac{x^2 - 6}{(x + 2)^2\,(2x - 1)} = \dfrac{A}{x + 2} + \dfrac{B}{(x + 2)^2} + \dfrac{C}{2x - 1}$

$x^2 - 6 = A\,(x + 2)\,(2x - 1) + B\,(2x - 1) + C\,(x + 2)^2$

$x = -2 : \quad -2 = -5B \qquad\qquad \Rightarrow B = \quad \frac{2}{5}$

$x = \quad \frac{1}{2} : -\frac{23}{4} = \frac{25}{4}C \qquad\qquad \Rightarrow C = -\frac{23}{25}$

$x = \quad 1 : \quad -5 = 3A + B + 9C \Rightarrow A = \quad \frac{24}{25}$

16) $\dfrac{4x^3 + 3x^2 + 5x - 2}{x^3\,(x + 2)} = \dfrac{A}{x} + \dfrac{B}{x^2} + \dfrac{C}{x^3} + \dfrac{D}{x + 2}$

$4x^3 + 3x^2 + 5x - 2 = Ax^2\,(x + 2) + Bx\,(x + 2) + C\,(x + 2) + Dx^3$

$x = -2 : -32 = -8D \Rightarrow D = \quad 4$

$x = \quad 0 : \quad -2 = \quad 2C \Rightarrow C = -1$

$x = -1 : \quad -8 = \quad A - \quad B + \quad C - D \quad (E1)$

$x = \quad 1 : \quad 10 = 3A + 3B + 3C + D \quad (E2)$

(continued on next page)

153

Exercises 8.4

Substituting the values for C and D into E1 and E2 yields

$$\begin{cases} A - B = -3 \\ 3A + 3B = 9 \end{cases} \Rightarrow \begin{cases} A - B = -3 \quad \text{(E3)} \\ A + B = 3 \quad \text{(E4)} \end{cases}$$

E3 + E4 $\Rightarrow$ 2A = 0 $\Rightarrow$ A = 0; B = 3

19) $\dfrac{9x^2 - 3x + 8}{x\,(x^2 + 2)} = \dfrac{A}{x} + \dfrac{Bx + C}{x^2 + 2}$

$9x^2 - 3x + 8 = A\,(x^2 + 2) + (Bx + C)\,x$

x = 0 : 8 = 2A $\Rightarrow$ A = 4

x = 1 : 14 = 3A + B + C   (E1)

x = -1 : 20 = 3A + B - C   (E2)

E1 - E2 $\Rightarrow$ -6 = 2C $\Rightarrow$ C = -3; B = 5

22) $\dfrac{3x^3 + 13x - 1}{(x^2 + 4)^2} = \dfrac{Ax + B}{x^2 + 4} + \dfrac{Cx + D}{(x^2 + 4)^2}$

$3x^3 + 13x - 1 = (Ax + B)\,(x^2 + 4) + Cx + D$

$\qquad\qquad\qquad = Ax^3 + Bx^2 + (4A + C)\,x + (4B + D)$

Equating coefficients, we have the following :

x^3 : A = 3

x^2 : B = 0

x   : 4A + C = 13 $\Rightarrow$ C = 1

constant : 4B + D = -1 $\Rightarrow$ D = -1

25) By first dividing and then factoring, we have the following :

$2x + 3 + \dfrac{x + 5}{(2x + 1)\,(x - 1)} = 2x + 3 + \dfrac{A}{2x + 1} + \dfrac{B}{x - 1}$

$x + 5 = A\,(x - 1) + B\,(2x + 1)$

x = 1 : 6 = 3B $\Rightarrow$ B = 2

$x = -\tfrac{1}{2} : \tfrac{9}{2} = -\tfrac{3}{2}A \Rightarrow A = -3$

154

The answers for Exercises 1-7 are in the order A+B; A−B; 2A; −3B.

1) $\begin{bmatrix} 9 & -1 \\ -2 & 5 \end{bmatrix}$; $\begin{bmatrix} 1 & -3 \\ 4 & 1 \end{bmatrix}$; $\begin{bmatrix} 10 & -4 \\ 2 & 6 \end{bmatrix}$; $\begin{bmatrix} -12 & -3 \\ 9 & -6 \end{bmatrix}$

4) $\begin{bmatrix} 8 & 2 & 7 \\ 5 & 5 & 1 \end{bmatrix}$; $\begin{bmatrix} -8 & -6 & 7 \\ 5 & 3 & -7 \end{bmatrix}$; $\begin{bmatrix} 0 & -4 & 14 \\ 10 & 8 & -6 \end{bmatrix}$; $\begin{bmatrix} -24 & -12 & 0 \\ 0 & -3 & -12 \end{bmatrix}$

7) $\begin{bmatrix} -3 & 4 & 1 & 6 \\ 3 & 2 & 7 & -7 \end{bmatrix}$; $\begin{bmatrix} 3 & 4 & -1 & 0 \\ -1 & 2 & -7 & -3 \end{bmatrix}$; $\begin{bmatrix} 0 & 8 & 0 & 6 \\ 2 & 4 & 0 & -10 \end{bmatrix}$;

$\begin{bmatrix} 9 & 0 & -3 & -9 \\ -6 & 0 & -21 & 6 \end{bmatrix}$

The answers for Exercises 10-16 are in the order AB; BA.

10) $\begin{bmatrix} 4 & -2 \\ -2 & 1 \end{bmatrix}\begin{bmatrix} 2 & 1 \\ 4 & 2 \end{bmatrix} = \begin{bmatrix} 8-8 & 4-4 \\ -4+4 & -2+2 \end{bmatrix} = \begin{bmatrix} 0 & 0 \\ 0 & 0 \end{bmatrix}$;

$\begin{bmatrix} 2 & 1 \\ 4 & 2 \end{bmatrix}\begin{bmatrix} 4 & -2 \\ -2 & 1 \end{bmatrix} = \begin{bmatrix} 8-2 & -4+1 \\ 16-4 & -8+2 \end{bmatrix} = \begin{bmatrix} 6 & -3 \\ 12 & -6 \end{bmatrix}$

13) $\begin{bmatrix} 4 & -3 & 1 \\ -5 & 2 & 2 \end{bmatrix}\begin{bmatrix} 2 & 1 \\ 0 & 1 \\ -4 & 7 \end{bmatrix} = \begin{bmatrix} 8+0-4 & 4-3+7 \\ -10+0-8 & -5+2+14 \end{bmatrix} =$

$\begin{bmatrix} 4 & 8 \\ -18 & 11 \end{bmatrix}$; $\qquad \begin{bmatrix} 2 & 1 \\ 0 & 1 \\ -4 & 7 \end{bmatrix}\begin{bmatrix} 4 & -3 & 1 \\ -5 & 2 & 2 \end{bmatrix} =$

(continued on next page)

$$\begin{bmatrix} 8-5 & -6+2 & 2+2 \\ 0-5 & 0+2 & 0+2 \\ -16-35 & 12+14 & -4+14 \end{bmatrix} = \begin{bmatrix} 3 & -4 & 4 \\ -5 & 2 & 2 \\ -51 & 26 & 10 \end{bmatrix}$$

16) $\begin{bmatrix} 1 & 2 & 3 \\ 2 & 3 & 1 \\ 3 & 1 & 2 \end{bmatrix} \begin{bmatrix} 2 & 0 & 0 \\ 0 & 2 & 0 \\ 0 & 0 & 2 \end{bmatrix} = \begin{bmatrix} 2+0+0 & 0+4+0 & 0+0+6 \\ 4+0+0 & 0+6+0 & 0+0+2 \\ 6+0+0 & 0+2+0 & 0+0+4 \end{bmatrix} =$

$$\begin{bmatrix} 2 & 4 & 6 \\ 4 & 6 & 2 \\ 6 & 2 & 4 \end{bmatrix}; \qquad \begin{bmatrix} 2 & 0 & 0 \\ 0 & 2 & 0 \\ 0 & 0 & 2 \end{bmatrix} \begin{bmatrix} 1 & 2 & 3 \\ 2 & 3 & 1 \\ 3 & 1 & 2 \end{bmatrix} =$$

$$\begin{bmatrix} 2+0+0 & 4+0+0 & 6+0+0 \\ 0+4+0 & 0+6+0 & 0+2+0 \\ 0+0+6 & 0+0+2 & 0+0+4 \end{bmatrix} = \begin{bmatrix} 2 & 4 & 6 \\ 4 & 6 & 2 \\ 6 & 2 & 4 \end{bmatrix}$$

19) $\begin{bmatrix} 4 & -2 \\ 0 & 3 \\ -7 & 5 \end{bmatrix} \begin{bmatrix} 3 \\ 4 \end{bmatrix} = \begin{bmatrix} 12-8 \\ 0+12 \\ -21+20 \end{bmatrix} = \begin{bmatrix} 4 \\ 12 \\ -1 \end{bmatrix}$

22) $\begin{bmatrix} 1 & 2 & -3 \\ 4 & -5 & 6 \end{bmatrix} \begin{bmatrix} 1 & -1 & 0 & 2 \\ -2 & 3 & 1 & 0 \\ 0 & 4 & 0 & -3 \end{bmatrix} =$

$$\begin{bmatrix} (1-4+0) & (-1+6-12) & (0+2+0) & (2+0+9) \\ (4+10+0) & (-4-15+24) & (0-5+0) & (8+0-18) \end{bmatrix} =$$

(continued on next page)

$$\begin{bmatrix} -3 & -7 & 2 & 11 \\ 14 & 5 & -5 & -10 \end{bmatrix}$$

25) $A(B+C) = \begin{bmatrix} 1 & 2 \\ 0 & -3 \end{bmatrix}\begin{bmatrix} 5 & 0 \\ 1 & 1 \end{bmatrix} = \begin{bmatrix} 7 & 2 \\ -3 & -3 \end{bmatrix};$

$AB + AC = \begin{bmatrix} 8 & 1 \\ -9 & -3 \end{bmatrix} + \begin{bmatrix} -1 & 1 \\ 6 & 0 \end{bmatrix} = \begin{bmatrix} 7 & 2 \\ -3 & -3 \end{bmatrix}$

28) $(c+d)A = (c+d)\begin{bmatrix} a_{11} & a_{12} \\ a_{21} & a_{22} \end{bmatrix}$

$= \begin{bmatrix} (c+d)a_{11} & (c+d)a_{12} \\ (c+d)a_{21} & (c+d)a_{22} \end{bmatrix}$

$= \begin{bmatrix} ca_{11}+da_{11} & ca_{12}+da_{12} \\ ca_{21}+da_{21} & ca_{22}+da_{22} \end{bmatrix}$

$= \begin{bmatrix} ca_{11} & ca_{12} \\ ca_{21} & ca_{22} \end{bmatrix} + \begin{bmatrix} da_{11} & da_{12} \\ da_{21} & da_{22} \end{bmatrix}$

$= c\begin{bmatrix} a_{11} & a_{12} \\ a_{21} & a_{22} \end{bmatrix} + d\begin{bmatrix} a_{11} & a_{12} \\ a_{21} & a_{22} \end{bmatrix}$

$= cA + dA$

31) $\begin{bmatrix} 2 & -4 & | & 1 & 0 \\ 1 & 3 & | & 0 & 1 \end{bmatrix} \xrightarrow{R_1 - R_2} \begin{bmatrix} 1 & -7 & | & 1 & -1 \\ 1 & 3 & | & 0 & 0 \end{bmatrix} R_2 - R_1$

$$\begin{bmatrix} 1 & -7 & | & 1 & -1 \\ 0 & 10 & | & -1 & 2 \end{bmatrix} \frac{1}{10}R_2,\quad \begin{bmatrix} 1 & -7 & | & 1 & -1 \\ 0 & 1 & | & -\frac{1}{10} & \frac{2}{10} \end{bmatrix} R_1 + 7R_2$$

$$\begin{bmatrix} 1 & 0 & | & \frac{3}{10} & \frac{4}{10} \\ 0 & 1 & | & -\frac{1}{10} & \frac{2}{10} \end{bmatrix};\quad A^{-1} = \frac{1}{10}\begin{bmatrix} 3 & 4 \\ -1 & 2 \end{bmatrix}\ \text{where A is the original matrix}$$

34) $$\begin{bmatrix} 3 & -1 & | & 1 & 0 \\ 6 & -2 & | & 0 & 1 \end{bmatrix} R_2 - 2R_1,\quad \begin{bmatrix} 3 & -1 & | & 1 & 0 \\ 0 & 0 & | & -2 & 1 \end{bmatrix}$$

Since the identity cannot be obtained on the left, <u>no inverse exists.</u>

37) $$\begin{bmatrix} -2 & 2 & 3 & | & 1 & 0 & 0 \\ 1 & -1 & 0 & | & 0 & 1 & 0 \\ 0 & 1 & 4 & | & 0 & 0 & 1 \end{bmatrix} R_1 + 2R_2 \leftrightarrow R_2$$

$$\begin{bmatrix} 1 & -1 & 0 & | & 0 & 1 & 0 \\ 0 & 0 & 3 & | & 1 & 2 & 0 \\ 0 & 1 & 4 & | & 0 & 0 & 1 \end{bmatrix} R_1 + R_3$$

$$\begin{bmatrix} 1 & 0 & 4 & | & 0 & 1 & 1 \\ 0 & 0 & 3 & | & 1 & 2 & 0 \\ 0 & 1 & 4 & | & 0 & 0 & 1 \end{bmatrix} \frac{1}{3}R_2 \leftrightarrow R_3$$

$$\begin{bmatrix} 1 & 0 & 4 & | & 0 & 1 & 1 \\ 0 & 1 & 4 & | & 0 & 0 & 1 \\ 0 & 0 & 1 & | & \frac{1}{3} & \frac{2}{3} & 0 \end{bmatrix} \begin{matrix} R_1 - 4R_3 \\ R_2 - 4R_3 \end{matrix}$$

(continued on next page)

$$\begin{bmatrix} 1 & 0 & 0 & | & -\frac{4}{3} & -\frac{5}{3} & 1 \\ 0 & 1 & 0 & | & -\frac{4}{3} & -\frac{8}{3} & 1 \\ 0 & 0 & 1 & | & \frac{1}{3} & \frac{2}{3} & 0 \end{bmatrix};$$ $A^{-1} = \frac{1}{3} \begin{bmatrix} -4 & -5 & 3 \\ -4 & -8 & 3 \\ 1 & 2 & 0 \end{bmatrix}$ where A is the original matrix

40) $\begin{bmatrix} 1 & 1 & 1 & | & 1 & 0 & 0 \\ 2 & 2 & 2 & | & 0 & 1 & 0 \\ 3 & 3 & 3 & | & 0 & 0 & 1 \end{bmatrix} \begin{array}{c} \\ R_2 - 2R_1 \\ R_3 - 3R_1 \end{array} \begin{bmatrix} 1 & 1 & 1 & | & 1 & 0 & 0 \\ 0 & 0 & 0 & | & -2 & 1 & 0 \\ 0 & 0 & 0 & | & -3 & 0 & 1 \end{bmatrix};$

Since the identity cannot be obtained on the left, <u>no inverse exists.</u>

43) $\begin{bmatrix} a & 0 & | & 1 & 0 \\ 0 & b & | & 0 & 1 \end{bmatrix} \begin{array}{c} (1/a)\,R_1 \\ \underline{(1/b)\,R_2} \end{array} \begin{bmatrix} 1 & 0 & | & a^{-1} & 0 \\ 0 & 1 & | & 0 & b^{-1} \end{bmatrix}$

The inverse is the matrix with main diagonal elements a^{-1} and b^{-1}. The required conditions are that a and b are nonzero to avoid division by zero.

45)
$$AI_3 = \begin{bmatrix} a_{11} & a_{12} & a_{13} \\ a_{21} & a_{22} & a_{23} \\ a_{31} & a_{32} & a_{33} \end{bmatrix} \begin{bmatrix} 1 & 0 & 0 \\ 0 & 1 & 0 \\ 0 & 0 & 1 \end{bmatrix} = \begin{bmatrix} a_{11} & a_{12} & a_{13} \\ a_{21} & a_{22} & a_{23} \\ a_{31} & a_{32} & a_{33} \end{bmatrix} = A$$

$$I_3A = \begin{bmatrix} 1 & 0 & 0 \\ 0 & 1 & 0 \\ 0 & 0 & 1 \end{bmatrix} \begin{bmatrix} a_{11} & a_{12} & a_{13} \\ a_{21} & a_{22} & a_{23} \\ a_{31} & a_{32} & a_{33} \end{bmatrix} = \begin{bmatrix} a_{11} & a_{12} & a_{13} \\ a_{21} & a_{22} & a_{23} \\ a_{31} & a_{32} & a_{33} \end{bmatrix} = A$$

46) Show the same conditions as in Exercise 45 with a square matrix A of order 4.

49)
$$X = A^{-1}B = \frac{1}{3} \begin{bmatrix} -4 & -5 & 3 \\ -4 & -8 & 3 \\ 1 & 2 & 0 \end{bmatrix} \begin{bmatrix} 1 \\ 3 \\ -2 \end{bmatrix} = \frac{1}{3} \begin{bmatrix} -25 \\ -34 \\ 7 \end{bmatrix}; \quad \left(-\frac{25}{3}, -\frac{34}{3}, \frac{7}{3} \right)$$

1) $M_{11} = \begin{vmatrix} 3 & 2 \\ 7 & 0 \end{vmatrix} = -14 = A_{11};$ $M_{12} = \begin{vmatrix} 0 & 2 \\ -5 & 0 \end{vmatrix} = 10;$ $A_{12} = -10;$

   $M_{13} = \begin{vmatrix} 0 & 3 \\ -5 & 7 \end{vmatrix} = 15 = A_{13};$ $M_{21} = \begin{vmatrix} 4 & -1 \\ 7 & 0 \end{vmatrix} = 7;$ $A_{21} = -7;$

   $M_{22} = \begin{vmatrix} 2 & -1 \\ -5 & 0 \end{vmatrix} = -5 = A_{22};$ $M_{23} = \begin{vmatrix} 2 & 4 \\ -5 & 7 \end{vmatrix} = 34;$ $A_{23} = -34;$

   $M_{31} = \begin{vmatrix} 4 & -1 \\ 3 & 2 \end{vmatrix} = 11 = A_{31};$ $M_{32} = \begin{vmatrix} 2 & -1 \\ 0 & 2 \end{vmatrix} = 4;$ $A_{32} = -4;$

   $M_{33} = \begin{vmatrix} 2 & 4 \\ 0 & 3 \end{vmatrix} = 6 = A_{33}$

4) $M_{11} = 2 = A_{11};$ $M_{12} = 3$ and $A_{12} = -3;$ $M_{21} = 4$ and $A_{21} = -4;$
   $M_{22} = -6 = A_{22}$

Let the matrix in Exercises 13, 16, and 19 be called A.

7) $\begin{vmatrix} 7 & -1 \\ 5 & 0 \end{vmatrix} = (7)\,(0) - (-1)\,(5) = 0 + 5 = 5$

10) $\begin{vmatrix} 6 & 4 \\ -3 & 2 \end{vmatrix} = 12 - (-12) = 24$

13) Expand along the first row.
   $|A| = a_{11}A_{11} + a_{12}A_{12} + a_{13}A_{13} = 3\,(-17) + 1\,(-26) - 2\,(24) = -125$

16) Expand along the second row.
   $|A| = a_{21}A_{21} + a_{22}A_{22} + a_{23}A_{23} = 1\,(17) + 0\,(A_{22}) + 4\,(30) = 137$

19) Expand along the first row.

$$|A| = -b \begin{vmatrix} 0 & c & 0 \\ a & 0 & 0 \\ 0 & 0 & d \end{vmatrix} ; \text{ Expand again along the first row.}$$

$$(-b)(-c) \begin{vmatrix} a & 0 \\ 0 & d \end{vmatrix} = abcd$$

LS denotes Left Side and RS denotes Right Side for Exercises 22-28.

22) LS = ad − bc;   RS = −(bc − ad) = ad − bc

25) LS = ad − bc;   RS = akb + ad − akb − bc = ad − bc

28) LS = ad − bc + af − be; RS = ad + af − bc − be

31) (a) $f(x) = \begin{vmatrix} 1 - x & 2 \\ 3 & 2 - x \end{vmatrix} = x^2 - 3x - 4$

   (b) $(x - 4)(x + 1) = 0 \Rightarrow x = -1, 4$

34) (a) $f(x) = \begin{vmatrix} 2 - x & -4 \\ -3 & 5 - x \end{vmatrix} = x^2 - 7x - 2$

   (b) $x^2 - 7x - 2 = 0 \Rightarrow x = \dfrac{7 \pm \sqrt{49 + 8}}{2} = \dfrac{7 \pm \sqrt{57}}{2}$

37) (a) $f(x) = \begin{vmatrix} 0 - x & 2 & -2 \\ -1 & 3 - x & 1 \\ -3 & 3 & 1 - x \end{vmatrix}$ Expand along the first row.

   $= (-x)\big((3 - x)(1 - x) - 3\big) - 2\big((x - 1) + 3\big) - 2\big(-3 + 3(3 - x)\big)$

   $= -x^3 + 4x^2 + 4x - 16 = (x - 2)(-x^2 + 2x + 8) = (x \pm 2)(-x + 4)$

   (b) $x = -2, 2, 4$

40) Expand the matrix along the first row to yield :

$$i \begin{vmatrix} -2 & 3 \\ 1 & -4 \end{vmatrix} - j \begin{vmatrix} 1 & 3 \\ 2 & -4 \end{vmatrix} + k \begin{vmatrix} 1 & -2 \\ 2 & 1 \end{vmatrix} = 5i + 10j + 5k$$

Exercises 8.7

1) The determinant value is negated since rows 2 & 3 are interchanged.

4) Row 1 is replaced by $R_1 - R_2$, there is no change in the det. value.

7) Row 1 and row 3 are identical making the determinant value zero.

10) −1 is factored out of row 1.

13) Column 3 is replaced by $C_3 + 2C_1$, there is no change in the determinant value.

16)
$$\begin{vmatrix} -3 & 0 & 4 \\ 1 & 2 & 0 \\ 4 & 1 & -1 \end{vmatrix} \xrightarrow{R_2 - 2R_3} = \begin{vmatrix} -3 & 0 & 4 \\ -7 & 0 & 2 \\ 4 & 1 & -1 \end{vmatrix} \text{Expand by the second column}$$

$$= (-1) \begin{vmatrix} -3 & 4 \\ -7 & 2 \end{vmatrix} = (-1)(22) = -22$$

19)
$$\begin{vmatrix} 2 & 2 & -3 \\ 3 & 6 & 9 \\ -2 & 5 & 4 \end{vmatrix} \text{3 is a common factor in row 2}$$

$$= (3) \begin{vmatrix} 2 & 2 & -3 \\ 1 & 2 & 3 \\ -2 & 5 & 4 \end{vmatrix} \begin{matrix} R_1 - 2R_2 \\ \\ R_3 + 2R_2 \end{matrix} \xrightarrow{\quad} = (3) \begin{vmatrix} 0 & -2 & -9 \\ 1 & 2 & 3 \\ 0 & 9 & 10 \end{vmatrix} \begin{matrix} \text{Expand by the} \\ \text{first column} \end{matrix}$$

$$= (3)(-1) \begin{vmatrix} -2 & -9 \\ 9 & 10 \end{vmatrix} = (-3)(61) = -183$$

22)
$$\begin{vmatrix} 3 & 2 & 0 & 4 \\ -2 & 0 & 5 & 0 \\ 4 & -3 & 1 & 6 \\ 2 & -1 & 2 & 0 \end{vmatrix} \begin{matrix} \\ R_2 - 5R_3 \\ \\ R_4 - 2R_3 \end{matrix} \xrightarrow{\quad} = \begin{vmatrix} 3 & 2 & 0 & 4 \\ -22 & 15 & 0 & -30 \\ 4 & -3 & 1 & 6 \\ -6 & 5 & 0 & -12 \end{vmatrix} \begin{matrix} \text{Expand by the} \\ \text{third column.} \end{matrix}$$

(continued on next page)

162

$$= (1) \begin{vmatrix} 3 & 2 & 4 \\ -22 & 15 & -30 \\ -6 & 5 & -12 \end{vmatrix} \begin{matrix} C_1 - C_2 \\ \\ \\ \end{matrix} = \begin{vmatrix} 1 & 2 & 4 \\ -37 & 15 & -30 \\ -11 & 5 & -12 \end{vmatrix} \begin{matrix} \\ R_2 + 37\,R_1 \\ R_3 + 11\,R_1 \end{matrix}$$

$$= \begin{vmatrix} 1 & 2 & 4 \\ 0 & 89 & 118 \\ 0 & 27 & 32 \end{vmatrix} \begin{matrix} \text{Expand by the} \\ \text{first column.} \end{matrix} = (1) \begin{vmatrix} 89 & 118 \\ 27 & 32 \end{vmatrix} = (1)(-338) = -338$$

25) Using the hint, $\begin{vmatrix} 1 & 1 & 1 \\ a & b & c \\ a^2 & b^2 & c^2 \end{vmatrix} \begin{matrix} C_1 - C_2 \\ \\ C_3 - C_2 \end{matrix} = \begin{vmatrix} 0 & 1 & 0 \\ a-b & b & c-b \\ a^2-b^2 & b^2 & c^2-b^2 \end{vmatrix}$

$a - b$ is a common factor of column 1 and $c - b$ is a common factor of column 3

$$= (a-b)(c-b) \begin{vmatrix} 0 & 1 & 0 \\ 1 & b & 1 \\ a+b & b^2 & c+b \end{vmatrix} \text{Expand by the first row.}$$

$$= (a-b)(c-b)(-1) \begin{vmatrix} 1 & 1 \\ a+b & c+b \end{vmatrix} \begin{matrix} = (a-b)(b-c)(c+b-a-b) \\ = (a-b)(b-c)(c-a) \end{matrix}$$

28) $\begin{vmatrix} a & b & 0 & 0 \\ c & d & 0 & 0 \\ 0 & 0 & e & f \\ 0 & 0 & g & h \end{vmatrix} \begin{matrix} \text{Expand} \\ \text{by the} \\ \text{first} \\ \text{column.} \end{matrix} = a \begin{vmatrix} d & 0 & 0 \\ 0 & e & f \\ 0 & g & h \end{vmatrix} - c \begin{vmatrix} b & 0 & 0 \\ 0 & e & f \\ 0 & g & h \end{vmatrix} \begin{matrix} \text{Expand both by} \\ \text{the first row.} \end{matrix}$

$$= ad \begin{vmatrix} e & f \\ g & h \end{vmatrix} - cb \begin{vmatrix} e & f \\ g & h \end{vmatrix} = (ad-bc) \begin{vmatrix} e & f \\ g & h \end{vmatrix} = \begin{vmatrix} a & b \\ c & d \end{vmatrix} \begin{vmatrix} e & f \\ g & h \end{vmatrix}$$

Exercises 8.7

31) $0 = \begin{vmatrix} x & y & 1 \\ x_1 & y_1 & 1 \\ x_2 & y_2 & 1 \end{vmatrix} \begin{matrix} R_1 - R_2 \\ \\ R_3 - R_2 \end{matrix}$

$\Rightarrow 0 = \begin{vmatrix} x - x_1 & y - y_1 & 0 \\ x_1 & y_1 & 1 \\ x_2 - x_1 & y_2 - y_1 & 0 \end{vmatrix}$ Expand by the third column.

$\Rightarrow 0 = \begin{vmatrix} x - x_1 & y - y_1 \\ x_2 - x_1 & y_2 - y_1 \end{vmatrix}$

$\Rightarrow 0 = (x - x_1)(y_2 - y_1) - (y - y_1)(x_2 - x_1)$

$\Rightarrow (x - x_1)(y_2 - y_1) = (y - y_1)(x_2 - x_1)$

$\Rightarrow y - y_1 = \dfrac{y_2 - y_1}{x_2 - x_1}(x - x_1)$, which is an equation of a line through

the points (x_1, y_1) and (x_2, y_2).

Exercises 8.8

In this section, the values of the determinants $|D_x|$, $|D_y|$, ... are given. To find the values of the variables x, y, ..., simply divide $|D_x|$, $|D_y|$, ... by $|D|$.

1) $|D_x| = -28$, $|D_y| = 14$, $|D| = -7$; x = 4, y = -2

4) $|D_x| = -53$, $|D_y| = -106$, $|D| = 53$; x = -1, y = -2

7) $|D_x| = 76$, $|D_y| = 28$, $|D| = 53$; $x = \frac{76}{53}$, $y = \frac{28}{53}$

10) $|D_t| = \frac{11}{24}$, $|D_v| = -\frac{19}{24}$, $|D| = \frac{31}{120}$; $t = \frac{55}{31}$, $v = -\frac{95}{31}$

13) $|D| = 0$ so Cramer's Rule does not apply.

16) $|D| = 0$ so Cramer's Rule does not apply.

19) $|D_x| = -60$, $|D_y| = -90$, $|D_z| = 30$, $|D| = -30$;

x = 2, y = 3, z = -1

22) $|D_x| = -12$, $|D_y| = -24$, $|D_z| = 0$, $|D| = -4$;

x = 3, y = 6, z = 0

25) $|D_x| = -14$, $|D_y| = -31$, $|D_z| = -1$, $|D| = -21$;

$x = \frac{2}{3}$, $y = \frac{31}{21}$, $z = \frac{1}{21}$

28) $|D_x| = -11$, $|D_y| = -11$, $|D_z| = 0$, $|D_w| = 0$, $|D| = -11$;

$x = 1$, $y = 1$, $z = 0$, $w = 0$

1) $3x - 2y = 6$ is a line with x-intercept 2 and y-intercept -3. Use a dashed line with "<". Substitute the point $(0, 0)$ into $3x - 2y < 6$. $3(0) - 2(0) < 6 \Rightarrow 0 < 6$. This is true so all points on the same side of the line as the test point, $(0, 0)$, are included in the solution. See Figure 1.

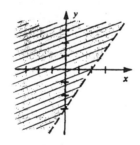
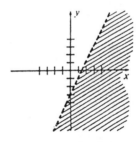

Figure 1   Figure 4

4) $2x - y = 3$ is a line with x-intercept $\frac{3}{2}$ and y-intercept -3. Use a dashed line. Test $(0, 0)$: $0 > 3$ is false so shade the side not containing $(0, 0)$. See Figure 4.

7) $x^2 + 1 = y$ is a parabola with y-intercept 1. Use a solid line. Test $(0, 0)$: $1 \leq 0$ is false so shade the side not containing $(0, 0)$. See Figure 7.

Exercises 8.9

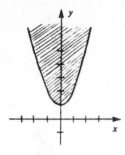

Figure 7

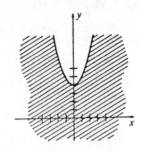

Figure 10

10) $x^2 + 4 = y$ is a parabola with y-intercept 4. Use a solid line. Test $(0, 0) : 4 \geq 0$ is true so shade the side containing $(0, 0)$. See Figure 10.

13) For $y - x < 0$, test $(1, 0) : -1 < 0$ is true so shade the area below $y - x = 0$. For $2x + 5y < 10$, test $(0, 0) : 0 < 10$ is true so shade the area below $2x + 5y = 10$. The final solution is the intersection of the two regions. See Figure 13.

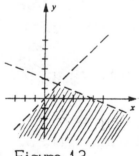

Figure 13

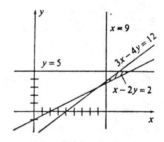

Figure 16

16) For $3x - 4y \geq 12$, test $(0, 0) : 0 \geq 12$ is false so shade the area below $3x - 4y = 12$. For $x - 2y \leq 2$, test $(0, 0) : 0 \leq 2$ is true so shade the area above $x - 2y = 2$. Shade the right side of $x = 9$ and below $y = 5$. The final solution is the intersection of all four regions. See Figure 16.

19) For $x^2 \le 1 - y$, test $(0, 0) : 0 \le 1$ is true so shade the inside of the parabola. For $x \ge 1 + y$, test $(0, 0) : 0 \ge 1$ is false so shade the area below $x = 1 + y$. The final solution is the intersection of the two regions. See Figure 19.

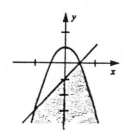

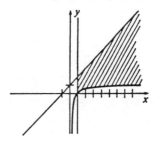

Figure 19   Figure 22

22) For $y \ge \log x$, test $(10, 0) : 0 \ge 1$ is false so shade the area above $y = \log x$. For $y - x \le 1$, test $(0, 0) : 0 \le 1$ is true so shade the area below $y - x = 1$. Shade the right side of $x = 1$. The final solution is the intersection of the 3 regions. See Figure 22.

25) Let x denote the number of sets of brand A.

Let y denote the number of sets of brand B.

$$\begin{cases} x \ge 2y \\ x \ge 20 \\ y \ge 10 \\ x + y \le 100 \end{cases}$$

The vertices of the solution region are $(20, 10)$, $(90, 10)$, and $(\frac{200}{3}, \frac{100}{3})$.

$x + y \le 100$ See Figure 25.

28) Let x denote the number of notebooks wholesaling for 55¢.

Let y denote the number of notebooks wholesaling for 85¢.

$$\begin{cases} .55x + .85y \le 600 \\ y \ge 300 \\ x \ge 400 \end{cases}$$

The vertices of the solution region are $(400, 300)$, $(400, \frac{7600}{17} \{ \approx 447.1 \})$, and $(\frac{6900}{11} \{ \approx 627.3 \}, 300)$. See Figure 28.

Exercises 8.9

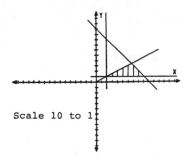

Scale 10 to 1

Figure 25

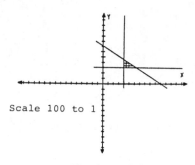

Scale 100 to 1

Figure 28

31) Let (x, y) denote the plant's location. A is $(0, 0)$; B is $(100, 0)$;

$$\begin{cases} 60^2 \leq \quad x^2 + y^2 \quad \leq 100^2 \\ 60^2 \leq (x - 100)^2 + y^2 \leq 100^2 \\ \quad y \geq 0 \end{cases}$$

The graph is the region in the first quadrant that lies between the two concentric circles with center $(0, 0)$ and radii 60 and 100, and also between the two concentric circles with center $(100, 0)$ and radii 60 and 100. Equating the different circle equations, we obtain the vertices of the solution region as $(50, 50 \sqrt{3})$, $(50, 10 \sqrt{11})$, $(18, 6 \sqrt{91})$, and $(82, 6 \sqrt{91})$. See Figure 31.

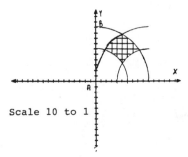

Scale 10 to 1

Figure 31

168

In this section, the constraining inequalities are given and then graphed. The vertex points { which the reader should verify } are summarized in table form with the resulting function values.

1) Let x denote the number of Set Point rackets.

Let y denote the number of Double Fault rackets.

Profit function : $P = 15x + 8y$

$$\begin{cases} 30 \leq y \leq 80 \\ 10 \leq x \leq 30 \\ x + y \leq 80 \end{cases}$$

(x, y)	P
A (10, 30)	390
B (30, 30)	690
C (30, 50)	850 ★
D (10, 70)	710

The maximum profit of $850 per day occurs when 30 Set Point rackets and 50 Double Fault rackets are manufactured. See Figure 1.

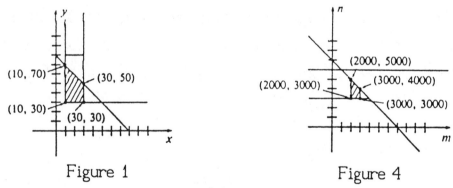

Figure 1   Figure 4

4) Let x denote the number of deluxe notebooks.

Let y denote the number of regular notebooks.

Difference function : $D = 0.25x + 0.15y$

(continued on next page)

$$\begin{cases} 2000 \le x \le 3000 \\ 3000 \le y \le 6000 \\ x + y \le 7000 \end{cases}$$

(x, y)	D
A (2000, 3000)	950
B (3000, 3000)	1200
C (3000, 4000)	1350 ★
E (2000, 5000)	1250

The maximum difference of $1350 occurs when 3000 deluxe notebooks and 4000 regular notebooks are produced. See Figure 4.

7) Let x denote the number of acres planted with alfalfa.

Let y denote the number of acres planted with corn.

Profit function : P = 110x − 4x − 20x + 150y − 6y − 10y

$$= 86x + 134y$$

seed cost $\quad\begin{cases} 4x + 6y \le 480 \end{cases}$

labor cost $\quad 20x + 10y \le 1400$

area $\quad x + y \le 100$

$\quad x, y \ge 0$

(x, y)	P
A (70, 0)	6020
B (45, 50)	10570
C (0, 80)	10720 ★
D (0, 0)	0

Note that the intersection points associated with the constraint x + y ≤ 100 do not fall in the solution region. The maximum profit of $10,720 occurs when 0 acres of alfalfa are planted and 80 acres of corn are planted. See Figure 7.

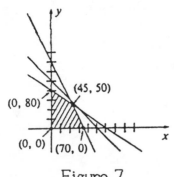

Figure 7

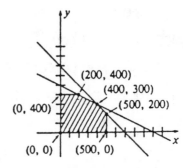

Figure 10

10) Let x denote the number of bags of peanuts.

Let y denote the number of bags of candy.

Profit function : $P = .15x + .20y$

$$\begin{array}{l} \underline{\text{purchase}} \\ \underline{\text{sell}} \end{array} \left\{ \begin{array}{l} .10x + .20y \leq 100 \\ x + y \leq 700 \\ 0 \leq x \leq 500 \\ 0 \leq y \leq 400 \end{array} \right.$$

(x, y)	P
A (500, 0)	75
B (500, 200)	115
C (400, 300)	120 ★
D (200, 400)	110
E (0, 400)	80
F (0, 0)	0

First inequality is $x + 2y \leq 1000$

The maximum profit of \$120 occurs when he sells 400 bags of peanuts and 300 bags of candy. See Figure 10.

13) Let x denote the number of trout.

Let y denote the number of bass.

Pound function : $P = 2x + 3y$

$$\begin{array}{l} \underline{\text{\# of fish}} \\ \underline{\text{cost}} \end{array} \left\{ \begin{array}{l} x + y \leq 5000 \\ .50x + .75y \leq 3000 \\ x, y \geq 0 \end{array} \right.$$

(x, y)	P
A (5000, 0)	10,000
B (3000, 2000)	12,000 ★
C (0, 4000)	12,000 ★
D (0, 0)	0

2nd inequality is $2x + 3y \leq 12000$

Points B and C yield 12,000 pounds of fish and so do all the points between B and C. The solution could be written as :

$\{(x, y) : 2x + 3y = 12,000, 0 \leq x \leq 3000,$ and x is an integer$\}$

See Figure 13.

Exercises 8.10

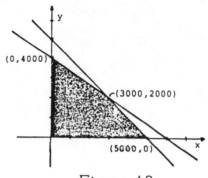

Figure 13

Exercises 8.11

In the Exercises, let Ei denote Equation i.

1) $4\,E1 + 3\,E2 \Rightarrow 23x = 19 \Rightarrow x = \frac{19}{23}$;

   $-5\,E1 + 2\,E2 \Rightarrow 23y = -18 \Rightarrow y = -\frac{18}{23}$

4) Solve E2 for x $(x = y + 7)$ and substitute into E1 to yield

   $y^2 + 7y + 12 = 0 \Rightarrow y = -3, -4$ and $x = 4, 3.$ (4, −3) and (3, −4)

7) $-4\,E1 + E2 \Rightarrow -14/y = -27 \Rightarrow y = \frac{14}{27}$;

   $2\,E1 + 3\,E2 \Rightarrow 14/x = 17 \Rightarrow x = \frac{14}{17}$; $(\frac{14}{17}, \frac{14}{27})$

10) Solve E1 for x and substitute into E3 to yield

$$\begin{cases} y - 5z = 3 & \text{(E4)} \\ -6y + z = -1 & \text{(E5)} \end{cases}$$

   $6\,E4 + E5 \Rightarrow -29z = 17 \Rightarrow z = -\frac{17}{29}$;

   $E4 + 5\,E5 \Rightarrow -29y = -2 \Rightarrow y = \frac{2}{29}$; $x = -\frac{6}{29}$ $(-\frac{6}{29}, \frac{2}{29}, -\frac{17}{29})$

13) $E1 - E2 \Rightarrow x - 5z = -1 \Rightarrow x = 5z - 1$; Substitute this value

   into E1 to obtain $y = \dfrac{-19z + 5}{2}$; $\left(5c - 1, \dfrac{-19c + 5}{2}, c\right)$ is the

   general solution where c is any real number.

16) Solve E1 for w and substitute into E2, E3, and E4 to obtain

$$\begin{cases} 5x + y + 2z = 10 & \text{(E5)} \\ -3x - y - 5z = 2 & \text{(E6)} \\ 5x - 2y + 13z = -9 & \text{(E7)} \end{cases}$$

Solve E6 for y and substitute into E5 and E7 to obtain

$$\begin{cases} 2x - 3z = 12 & \text{(E8)} \\ 11x + 23z = -3 & \text{(E9)} \end{cases}$$

$11\,E8 - 2\,E9 \Rightarrow -79z = 158 \Rightarrow z = -2 \qquad\qquad (3, -1, -2, 4)$

$23\,E8 + 3\,E9 \Rightarrow 79x = 237 \Rightarrow x = 3,\ y = -1,\ w = 4$

19) The solution is the intersection
of the following 3 regions :
above $x - 2y = 2$;
below $y - 3x = 4$;
below $2x + y = 4$;
See Figure 19.

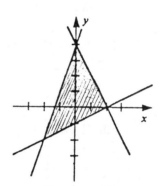

Figure 19

Let A denote each of the matrices in Exercises 21-30, 33-36.

22) $\begin{vmatrix} 3 & 4 \\ -6 & -5 \end{vmatrix} = -15 + 24 = 9$

25) Expand along R_1. $|A| = 2\,(-7) + 3\,(-5) + 5\,(-11) = -84$

28) $\begin{vmatrix} 1 & 2 & 0 & 3 & 1 \\ -2 & -1 & 4 & 1 & 2 \\ 3 & 0 & -1 & 0 & -1 \\ 2 & -3 & 2 & -4 & 2 \\ -1 & 1 & 0 & 1 & 3 \end{vmatrix} \begin{matrix} \\ R_2 + 4R_3 \\ \\ R_4 + 2R_3 \\ \\ \end{matrix} = \begin{vmatrix} 1 & 2 & 0 & 3 & 1 \\ 10 & -1 & 0 & 1 & -2 \\ 3 & 0 & -1 & 0 & -1 \\ 8 & -3 & 0 & -4 & 0 \\ -1 & 1 & 0 & 1 & 3 \end{vmatrix} =$

$$-1 \begin{vmatrix} 1 & 2 & 3 & 1 \\ 10 & -1 & 1 & -2 \\ 8 & -3 & -4 & 0 \\ -1 & 1 & 1 & 3 \end{vmatrix} \begin{matrix} \\ R_2 + 2R_1 \\ \\ R_4 - 3R_1 \end{matrix} = -1 \begin{vmatrix} 1 & 2 & 3 & 1 \\ 12 & 3 & 7 & 0 \\ 8 & -3 & -4 & 0 \\ -4 & -5 & -8 & 0 \end{vmatrix} =$$

Since 4 is a common factor of C_1 and -1 of R_3.

$$\begin{vmatrix} 12 & 3 & 7 \\ 8 & -3 & -4 \\ -4 & -5 & -8 \end{vmatrix} \downarrow = -4 \begin{vmatrix} 3 & 3 & 7 \\ 2 & -3 & -4 \\ 1 & 5 & 8 \end{vmatrix} \begin{matrix} R_1 - 3R_3 \\ R_2 - 2R_3 \\ \end{matrix} = -4 \begin{vmatrix} 0 & -12 & -17 \\ 0 & -13 & -20 \\ 1 & 5 & 8 \end{vmatrix}$$

$$= -4 \begin{vmatrix} -12 & -17 \\ -13 & -20 \end{vmatrix} = -4(19) = -76$$

31) This is an extension of Exercise 27 of Section 8.7. Expanding along R_1, only a_{11} is not 0. Expanding along the new R_1 again, only a_{22} is not 0. Repeating this process yields $|A| = a_{11}a_{22}...a_{nn}$, the product of the main diagonal elements.

34) $\begin{bmatrix} 2 & -1 & 0 & | & 1 & 0 & 0 \\ 1 & 4 & 2 & | & 0 & 1 & 0 \\ 3 & -2 & 1 & | & 0 & 0 & 1 \end{bmatrix} \begin{matrix} R_1 - 2R_2 \leftrightarrow R_2 \\ \\ R_3 - 3R_2 \end{matrix} \longrightarrow$

$\begin{bmatrix} 1 & 4 & 2 & | & 0 & 1 & 0 \\ 0 & -9 & -4 & | & 1 & -2 & 0 \\ 0 & -14 & -5 & | & 0 & -3 & 1 \end{bmatrix} \begin{matrix} \\ 3R_2 - 2R_3 \\ \\ \end{matrix} \longrightarrow$

$\begin{bmatrix} 1 & 4 & 2 & | & 0 & 1 & 0 \\ 0 & 1 & -2 & | & 3 & 0 & -2 \\ 0 & -14 & -5 & | & 0 & -3 & 1 \end{bmatrix} \begin{matrix} R_1 - 4R_2 \\ \\ R_3 + 14R_2 \end{matrix} \longrightarrow$

$$\left[\begin{array}{ccc|ccc} 1 & 0 & 10 & -12 & 1 & 8 \\ 0 & 1 & -2 & 3 & 0 & -2 \\ 0 & 0 & -33 & 42 & -3 & -27 \end{array}\right] -\tfrac{1}{33} R_3$$

$$\left[\begin{array}{ccc|ccc} 1 & 0 & 10 & -12 & 1 & 8 \\ 0 & 1 & -2 & 3 & 0 & -2 \\ 0 & 0 & 1 & -\tfrac{14}{11} & \tfrac{1}{11} & \tfrac{9}{11} \end{array}\right] \begin{array}{l} R_1 - 10\,R_3 \\ R_2 + 2\,R_3 \end{array}$$

$$\left[\begin{array}{ccc|ccc} 1 & 0 & 0 & \tfrac{8}{11} & \tfrac{1}{11} & -\tfrac{2}{11} \\ 0 & 1 & 0 & \tfrac{5}{11} & \tfrac{2}{11} & -\tfrac{4}{11} \\ 0 & 0 & 1 & -\tfrac{14}{11} & \tfrac{1}{11} & \tfrac{9}{11} \end{array}\right] ; \quad A^{-1} = \tfrac{1}{11} \left[\begin{array}{ccc} 8 & 1 & -2 \\ 5 & 2 & -4 \\ -14 & 1 & 9 \end{array}\right]$$

37) $\left[\begin{array}{ccc} 4 & -5 & 6 \\ 4 & -11 & 5 \end{array}\right]$

40) $\left[\begin{array}{cc} 0 & -37 \\ 15 & -6 \end{array}\right]$

43) $\left[\begin{array}{cc} a & 3a \\ 2b & 4b \end{array}\right]$

46) A matrix of order 3 multiplied by its inverse is equal to the identity matrix of order 3. $A\,A^{-1} = I_3$

49) $\left|\begin{array}{cc} 2 - x & 3 \\ 1 & -4 - x \end{array}\right| = 0 \Rightarrow (2 - x)(-4 - x) - 3 = 0$

$$\Rightarrow x^2 + 2x - 11 = 0$$

$$\Rightarrow x = \frac{-2 \pm \sqrt{4 + 44}}{2} = -1 \pm 2\sqrt{3}$$

52) $\begin{bmatrix} a_{11} & a_{12} \\ a_{21} & a_{22} \end{bmatrix} \begin{array}{|cc} 1 & 0 \\ 0 & 1 \end{array}$ $\dfrac{1/a_{11}\,R_1}{\xrightarrow{\hspace{1cm}}}$ (Note : a_{11} or a_{21} must be nonzero since $|A| \neq 0$. The rows could be swapped.)

$\begin{bmatrix} 1 & a_{12}/a_{11} \\ a_{21} & a_{22} \end{bmatrix} \begin{array}{|cc} 1/a_{11} & 0 \\ 0 & 1 \end{array}$ $\underline{R_2 - a_{21}R_1}$

$\begin{bmatrix} 1 & a_{12}/a_{11} \\ 0 & (a_{11}a_{22} - a_{12}a_{21})/a_{11} \end{bmatrix} \begin{array}{|cc} 1/a_{11} & 0 \\ -a_{21}/a_{11} & 1 \end{array}$ Let $|A| = a_{11}a_{22} - a_{12}a_{21}$. $\underline{a_{11}/|A|\,R_2}$

$\begin{bmatrix} 1 & a_{12}/a_{11} \\ 0 & 1 \end{bmatrix} \begin{array}{|cc} 1/a_{11} & 0 \\ -a_{21}/|A| & a_{11}/|A| \end{array}$ $\dfrac{R_1 - (a_{12}/a_{11})\,R_2}{\xrightarrow{\hspace{2cm}}}$

$\begin{bmatrix} 1 & 0 \\ 0 & 1 \end{bmatrix} \begin{array}{|cc} 1/a_{11} + (a_{12}a_{21})/a_{11}|A| & -a_{12}/|A| \\ -a_{21}/|A| & a_{11}/|A| \end{array}$ $=$ (Use the definition of $|A|$ above.)

$\begin{bmatrix} 1 & 0 \\ 0 & 1 \end{bmatrix} \begin{array}{|cc} a_{22}/|A| & -a_{12}/|A| \\ -a_{21}/|A| & a_{11}/|A| \end{array}$; $A^{-1} = \dfrac{1}{|A|} \begin{bmatrix} a_{22} & -a_{12} \\ -a_{21} & a_{11} \end{bmatrix}$

55) Let x and y denote the length and width, respectively, of the rectangle.

$\underline{\text{area}}$ $\left\{ \begin{array}{l} xy = 4000 \quad \text{(E1)} \\ x^2 + y^2 = 100^2 \quad \text{(E2)} \end{array} \right.$ Solve E1 for y $(= 4000/x)$ and

$\underline{\text{diagonal}}$ substitute into E2.

$x^2 + \dfrac{4000^2}{x^2} = 100^2 \Rightarrow x^4 - 10{,}000x^2 + 16{,}000{,}000 = 0$

$\Rightarrow (x^2 - 2000)(x^2 - 8000) = 0$

$\Rightarrow x = 20\sqrt{5},\ 40\sqrt{5}$ and $y = 40\sqrt{5},\ 20\sqrt{5}$;

The dimensions are $20\sqrt{5}$ feet by $40\sqrt{5}$ feet.

58) Let r_1 denote the inside radius and r_2 the outside radius.

Inside distance = .90 (outside distance)

$\Rightarrow \qquad 2\Pi r_1 = .90\ (2\Pi r_2)$

$\Rightarrow \qquad r_1 = .90\ (r_1 + 10)$ (since $r_2 = r_1 + 10$)

$\Rightarrow \qquad .1r_1 = 9 \Rightarrow r_1 = 90$ feet and $r_2 = 100$ feet

61) Let x denote the number of lawn mowers produced.

Let y denote the number of edgers produced.

Profit function : $P = 100x + 80y$

machining $\left(\begin{array}{l} 6x + 4y \le 600 \\ \end{array}\right.$

welding $\quad 2x + 3y \le 300$

assembly $\quad 5x + 5y \le 550$

$\left. \qquad\quad x,\ y \ge 0 \right.$

(x, y)		P
A (100, 0)		10,000
B (80, 30)		10,400 ⋆
C (30, 80)		9,400
D (0, 100)		8,000
E (0, 0)		0

The maximum weekly profit of $10,400 occurs when 80 lawn mowers and 30 edgers are produced.

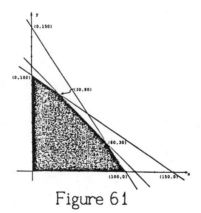

Figure 61

Exercises 9.1

P_n is the statement in the text for Exercises 1-22.

1) P_1 is true since $2(1) = 1(1 + 1) = 2$; Assume P_k is true :

$$2 + 4 + 6 + \cdots + 2k \qquad\qquad = k(k + 1)$$
$$2 + 4 + 6 + \cdots + 2k + 2(k + 1) = k(k + 1) + 2(k + 1)$$
$$= (k + 1)(k + 2)$$
$$= (k + 1)(k + 1 + 1)$$

Thus, P_{k+1} is true and the proof by math induction is complete.

4) P_1 is true since $6(1) - 3 = 3(1)^2 = 3$; Assume P_k is true :

$$3 + 9 + 15 + \cdots + 6k - 3 \qquad\qquad = 3k^2$$
$$3 + 9 + 15 + \cdots + 6k - 3 + 6(k + 1) - 3 = 3k^2 + 6(k + 1) - 3$$
$$= 3k^2 + 6k + 3$$
$$= 3(k^2 + 2k + 1)$$
$$= 3(k + 1)^2$$

Thus, P_{k+1} is true and the proof by math induction is complete.

7) P_1 is true since $1 \cdot 2^{1-1} = 1 + (1 - 1) \cdot 2^1 = 1$; Assume P_k is true :

$$1 + 2 \cdot 2 + 3 \cdot 2^2 + \cdots + k \cdot 2^{k-1} \qquad\qquad = 1 + (k - 1) \cdot 2^k$$
$$1 + 2 \cdot 2 + 3 \cdot 2^2 + \cdots + k \cdot 2^{k-1} + (k + 1) \cdot 2^k = 1 + (k - 1) \cdot 2^k$$
$$+ (k + 1) \cdot 2^k$$
$$= 1 + k \cdot 2^k - 2^k + k \cdot 2^k + 2^k$$
$$= 1 + k \cdot 2^1 \cdot 2^k$$
$$= 1 + ((k + 1) - 1) \cdot 2^{k+1}$$

Thus, P_{k+1} is true and the proof by math induction is complete.

10) P_1 is true since $1^3 = \left[\dfrac{1\,(1+1)}{2}\right]^2 = 1$; Assume P_k is true :

$$1^3 + 2^3 + 3^3 + \cdots + k^3 \qquad = \left[\dfrac{k\,(k+1)}{2}\right]^2$$

$$1^3 + 2^3 + 3^3 + \cdots + k^3 + (k+1)^3 = \left[\dfrac{k\,(k+1)}{2}\right]^2 + (k+1)^3$$

$$= \dfrac{(k+1)^2}{2^2}\left[(k)^2 + 4\,(k+1)\right]$$

$$= \dfrac{(k+1)^2}{2^2}\,(k+2)^2$$

$$= \left[\dfrac{(k+1)\,((k+1)+1)}{2}\right]^2$$

Thus, P_{k+1} is true and the proof by math induction is complete.

13) P_1 is true since $3^1 = \tfrac{3}{2}\,(3^1 - 1) = 3$; Assume P_k is true :

$$3 + 3^2 + 3^3 + \cdots + 3^k \qquad = \tfrac{3}{2}\,(3^k - 1)$$

$$3 + 3^2 + 3^3 + \cdots + 3^k + 3^{k+1} = \tfrac{3}{2}\,(3^k - 1) + 3^{k+1}$$

$$= \tfrac{3}{2}\cdot 3^k - \tfrac{3}{2} + 3\cdot 3^k$$

$$= \tfrac{9}{2}\cdot 3^k - \tfrac{3}{2}$$

$$= \tfrac{3}{2}\,(3\cdot 3^k - 1)$$

$$= \tfrac{3}{2}\,(3^{k+1} - 1)$$

Thus, P_{k+1} is true and the proof by math induction is complete.

16) P_1 is true since $1 + 2\,(1) \le 3^1$; Assume P_k is true : $1 + 2k \le 3^k$;
$1 + 2\,(k + 1) = 2k + 3 < 6k + 3$ which is $3\,(1 + 2k)$; Now

$3\,(1 + 2k) < 3\,(3^k)$ {from P_k} $= 3^{k+1}$; i.e. $1 + 2\,(k + 1) \le 3^{k+1}$;

Thus, P_{k+1} is true and the proof by math induction is complete.

19) For n = 1, $n^3 - n + 3 = 3$ and 3 is a factor of 3. Assume 3 is a factor of $k^3 - k + 3$. Now the (k + 1)st term is

$(k + 1)^3 - (k + 1) + 3 = k^3 + 3k^2 + 2k + 3$

$$= (k^3 - k + 3) + 3k^2 + 3k$$
$$= (k^3 - k + 3) + 3(k^2 + k)$$

By hypothesis, 3 is a factor of $k^3 - k + 3$ and 3 is a factor of $3(k^2 + k)$, so 3 is a factor of the (k + 1)st term. Thus, P_{k+1} is true and the proof by math induction is complete.

22) For n = 1, $10^{n+1} + 3 \cdot 10^n + 5 = 135$ and 9 is a factor of 135.

Assume 9 is a factor of $10^{k+1} + 3 \cdot 10^k + 5$. Now the (k + 1)st

term is $10^{k+2} + 3 \cdot 10^{k+1} + 5 = 10 \cdot 10^{k+1} + 10 \cdot 3 \cdot 10^k + 5$

$$= 10^{k+1} + 9 \cdot 10^{k+1} + 3 \cdot 10^k + 9 \cdot 3 \cdot 10^k + 5$$
$$= (10^{k+1} + 3 \cdot 10^k + 5) + 9(10^{k+1} + 3 \cdot 10^k)$$

By hypothesis, 9 is a factor of $10^{k+1} + 3 \cdot 10^k + 5$ and 9 is a factor

of $9(10^{k+1} + 3 \cdot 10^k)$, so 9 is a factor of the (k + 1)st term. Thus,

P_{k+1} is true and the proof by math induction is complete.

25) For n = 1, a - b is a factor of $a^1 - b^1$. Assume a - b is a factor of $a^k - b^k$. Following the hint for the (k + 1)st term,

$a^{k+1} - b^{k+1} = a^k \cdot a - b \cdot a^k + b \cdot a^k - b^k \cdot b$

$$= a^k(a - b) + (a^k - b^k) b.$$

Now (a - b) is a factor of $a^k(a - b)$ and by hypothesis, a - b is a

factor of $(a^k - b^k) b$ and therefore is a factor of the (k + 1)st term.

Thus, P_{k+1} is true and the proof by math induction is complete.

28) For n = 3, (n − 2)·180° = 180° which is true for any triangle,
thus P_3 is true. Assume the sum of the interior angles of a polygon
of k sides is (k − 2)·180°. Now any (k+1)-sided polygon can be
dissected into a k-sided polygon and a triangle by drawing a line
from vertex (i) to vertex (i+2). Its angles add up to (k − 2)·180°
(since it is k-sided, by hypothesis) + 180° (for the triangle), which
is (k − 1)·180°. Thus, P_{k+1} is true and the proof by math induction
is complete.

1) 6! = (6) (5) (4) (3) (2) (1) = 720

4) 2! 0! = 2·1 = 2   {0! = 1 by definition}

7) $\binom{7}{3} = \dfrac{7!}{3!\ 4!} = \dfrac{7\cdot6\cdot5}{3\cdot2} = 35$

10) $\binom{4}{4} = \dfrac{4!}{4!\ 0!} = 1$

Summation notation is introduced in the next section, but it is convenient
to use it here for representing the binomial expansions. This is merely
a "shorthand" notation for The Binomial Theorem on pg. 449 of the text.

13) $(a + b)^6 = \displaystyle\sum_{k=0}^{6} \binom{6}{k} (a)^{6-k} (b)^k = \binom{6}{0}a^6b^0 + \binom{6}{1}a^5b^1 + \binom{6}{2}a^4b^2$

$\qquad\qquad + \binom{6}{3}a^3b^3 + \binom{6}{4}a^2b^4 + \binom{6}{5}a^1b^5 + \binom{6}{6}a^0b^6$

$\qquad\qquad = a^6 + 6a^5b + 15a^4b^2 + 20a^3b^3 + 15a^2b^4 + 6ab^5 + b^6$

16) $(a + b)^9 = \sum\limits_{k=0}^{9} \binom{9}{k} (a)^{9-k} (b)^k = \binom{9}{0} a^9 b^0 + \binom{9}{1} a^8 b^1 + \binom{9}{2} a^7 b^2 +$

$\binom{9}{3} a^6 b^3 + \binom{9}{4} a^5 b^4 + \binom{9}{5} a^4 b^5 + \binom{9}{6} a^3 b^6 + \binom{9}{7} a^2 b^7 + \binom{9}{8} a^1 b^8 +$

$\binom{9}{9} a^0 b^9 = a^9 + 9a^8 b + 36a^7 b^2 + 84a^6 b^3 + 126a^5 b^4 + 126a^4 b^5 +$

$84a^3 b^6 + 36a^2 b^7 + 9ab^8 + b^9$

19) $(u^2 + 4v)^5 = \sum\limits_{k=0}^{5} \binom{5}{k} (u^2)^{5-k} (4v)^k = \binom{5}{0} (u^2)^5 (4v)^0 +$

$\binom{5}{1} (u^2)^4 (4v)^1 + \binom{5}{2} (u^2)^3 (4v)^2 + \binom{5}{3} (u^2)^2 (4v)^3 + \binom{5}{4} (u^2)^1 (4v)^4$

$\binom{5}{5} (u^2)^0 (4v)^5 = u^{10} + 20u^8 v + 160u^6 v^2 + 640u^4 v^3 + 1280u^2 v^4 +$

$1024v^5$

22) $(x^{1/2} - y^{-1/2})^6 = \sum\limits_{k=0}^{6} \binom{6}{k} (x^{1/2})^{6-k} (-y^{-1/2})^k = \binom{6}{0} (x^{1/2})^6 (-y^{-1/2})^0 +$

$\binom{6}{1} (x^{1/2})^5 (-y^{-1/2})^1 + \binom{6}{2} (x^{1/2})^4 (-y^{-1/2})^2 + \binom{6}{3} (x^{1/2})^3 (-y^{-1/2})^3 +$

$\binom{6}{4} (x^{1/2})^2 (-y^{-1/2})^4 + \binom{6}{5} (x^{1/2})^1 (-y^{-1/2})^5 + \binom{6}{6} (x^{1/2})^0 (-y^{-1/2})^6$

$= x^3 - 6x^{5/2} y^{-1/2} + 15x^2 y^{-1} - 20x^{3/2} y^{-3/2} + 15xy^{-2} - 6x^{1/2} y^{-5/2} + y^{-3}$

25) $\sum\limits_{k=0}^{3} \binom{25}{k} (3c^{2/5})^{25-k} (c^{4/5})^k = \binom{25}{0} (3c^{2/5})^{25} (c^{4/5})^0 +$

$\binom{25}{1} (3c^{2/5})^{24} (c^{4/5})^1 + \binom{25}{2} (3c^{2/5})^{23} (c^{4/5})^2 + \binom{25}{3} (3c^{2/5})^{22} (c^{4/5})^3$

$= (3^{25}) c^{10} + 25 (3^{24}) c^{52/5} + 300 (3^{23}) c^{54/5} + 2300 (3^{22}) c^{56/5}$

28) $\displaystyle\sum_{k=10}^{12} \binom{12}{k} (s)^{12-k} (-2t^3)^k = \binom{12}{10} s^2 (-2t^3)^{10} + \binom{12}{11} s^1 (-2t^3)^{11} +$

$\binom{12}{12} s^0 (-2t^3)^{12} = 66\,(2^{10})\,s^2 t^{30} - 12\,(2^{11})\,st^{33} + (2^{12})\,t^{36}$

For the following Exercises, the general formula for the kth term of

$(a + b)^n$ is $\binom{n}{k-1} (a)^{n-(k-1)} (b)^{k-1} = \binom{n}{k-1} (a)^{n-k+1} (b)^{k-1}$

31) $a = \frac{1}{2}u$, $b = -2v$, $n = 10$, and $k = 7$; Using the formula above, the

7th term is $\binom{10}{6} (\frac{1}{2}u)^4 (-2v)^6 = 210\,(\frac{1}{16}u^4)\,(64v^6) = 840u^4v^6$.

34) $a = rs$, $b = t$, $n = 7$, and $k = (4$ and $5)$; Using the formula above,

the 4th term is $\binom{7}{3} (rs)^4 (t)^3 = 35r^4s^4t^3$, and the 5th term is

$\binom{7}{4} (rs)^3 (t)^4 = 35r^3s^3t^4$.

37) Consider only the variable y in the expansion :

$(y^3)^{k-1} = y^6 \Rightarrow y^{3k-3} = y^6 \Rightarrow 3k - 3 = 6 \Rightarrow k = 3$

3rd term $= \binom{4}{2} (x)^2 (-2y^3)^2 = 24x^2y^6$

40) Consider only the variable y in the expansion :

$(y)^{8-k+1} (y^{-3})^{k-1} = y^0 \Rightarrow y^{12-4k} = y^0 \Rightarrow k = 3$

3rd term $= \binom{8}{2} (xy)^6 (-2y^{-3})^2 = 28x^6y^6 \cdot 4y^{-6} = 112x^6$

Exercises 9.3

For Exercises 1-16, the answers are listed in the order : a_1, a_2, a_3, a_4, a_5; and a_8. Simply substitute 1, 2, 3, 4, 5, and 8 for n in the formula for a_n to obtain the results.

1) 9, 6, 3, 0, -3; -12

4) 11, $\frac{21}{2}$, $\frac{31}{3}$, $\frac{41}{4}$, $\frac{51}{5}$; $\frac{81}{8}$

7) 1.9, 2.01, 1.999, 2.0001, 1.99999; 2.00000001

10) $-2\sqrt{2}$, $\frac{2}{3}\sqrt{3}$, 0, $-\frac{2}{5}\sqrt{5}$, $\frac{2}{3}\sqrt{6}$; $-\frac{10}{3}$

13) $\frac{2}{3}$, $\frac{2}{3}$, $\frac{8}{11}$, $\frac{8}{9}$, $\frac{32}{27}$; $\frac{128}{33}$

16) 0, 7, 26, 63, 124; 511

19) $a_1 = -3$; $a_2 = (a_1)^2 = (-3)^2 = 9$; $a_3 = (a_2)^2 = (9)^2 = 81$;

    $a_4 = (a_3)^2 = (81)^2 = (3^4)^2 = 3^8$; $a_5 = (a_4)^2 = (3^8)^2 = 3^{16}$

22) $a_1 = 3$; $a_2 = \dfrac{1}{a_1} = \dfrac{1}{3}$; $a_3 = \dfrac{1}{a_2} = \dfrac{1}{\frac{1}{3}} = 3$; $a_4 = \dfrac{1}{a_3} = \dfrac{1}{3}$; $a_5 = \dfrac{1}{a_4} = \dfrac{1}{\frac{1}{3}} = 3$

25) The desired sequence starts with 2n for n = 1, 2, 3, 4. To obtain

the fifth term of a, add $\dfrac{(n-1)\,(n-2)\,(n-3)\,(n-4)\,(a-10)}{24}$

which is 0 for n = 1, 2, 3, 4 and (a - 10) for n = 5.

$$a_n = 2n + \frac{(n-1)\,(n-2)\,(n-3)\,(n-4)\,(a-10)}{24}$$

Another possibility is $a_n = \begin{cases} 2n & \text{if } 1 \le n \le 4 \\ (n-4)\,a & \text{if } n \ge 5 \end{cases}$

28) 7 + 4 + 1 + (-2) + (-5) + (-8) = -3

31) 0 + (-1) + 0 + 3 + 8 + 15 = 25

34) $\frac{3}{2} + 1 + \frac{3}{4} + \frac{3}{5} + \frac{1}{2} + \frac{3}{7} = \frac{669}{140}$

37) 100 (100) = 10,000

40) $\displaystyle\sum_{k=1}^{n} (3k^2 - 2k + 1) = 3 \sum_{k=1}^{n} k^2 - 2 \sum_{k=1}^{n} k + \sum_{k=1}^{n} 1$

$$= 3 \left[\frac{n(n+1)(2n+1)}{6} \right] - 2 \left[\frac{n(n+1)}{2} \right] + n$$

$$= \frac{2n^3 + n^2 + n}{2}$$

43) Since the difference in terms is 4, the coefficient of the summation variable will be 4. Consider the term $4k + x$; when $k = 1$, $4k + x$ should be 1, therefore x must be -3.

$$\sum_{k=1}^{5} (4k - 3) \quad \text{or} \quad \sum_{k=0}^{4} (4k + 1)$$

46) $\displaystyle\sum_{k=1}^{4} \frac{k}{5k - 1}$ The numerators increase by 1, the denominators by 5.

49) $\displaystyle\sum_{k=1}^{7} (-1)^{k+1} \frac{1}{k}$ The terms alternate, the numerator is 1, and the denominators increase by 1.

52) $\displaystyle\sum_{k=1}^{98} \frac{1}{k(k+1)(k+2)}$ The denominator is the product of 3 consecutive integers. The first term of the 3 integers ranges from 1 to 98.

55) Proof of (i) by mathematical induction.

P_1 is true since $\displaystyle\sum_{k=1}^{1} (a_k + b_k) = a_1 + b_1 = \sum_{k=1}^{1} a_k + \sum_{k=1}^{1} b_k$;

Assume P_i is true : $\displaystyle\sum_{k=1}^{i} (a_k + b_k) = \sum_{k=1}^{i} a_k + \sum_{k=1}^{i} b_k$

$$\sum_{k=1}^{i+1} (a_k + b_k) = \sum_{k=1}^{i} (a_k + b_k) + (a_{i+1} + b_{i+1})$$

(continued on next page)

$$= \sum_{k=1}^{i} a_k + \sum_{k=1}^{i} b_k + (a_{i+1} + b_{i+1}) \quad \text{(By hypothesis)}$$

$$= \sum_{k=1}^{i} a_k + a_{i+1} + \sum_{k=1}^{i} b_k + b_{i+1}$$

$$= \sum_{k=1}^{i+1} a_k + \sum_{k=1}^{i+1} b_k$$

Thus, P_{i+1} is true and the proof by math induction is complete.

(ii) Proof of (ii) by mathematical induction. Prove part (iii) first, then use $c = -1$ in the proof of part (ii).

P_1 is true since $\displaystyle\sum_{k=1}^{1} a_k - b_k = a_1 - b_1 = a_1 + (-1) b_1$

$$= \sum_{k=1}^{1} a_k - \sum_{k=1}^{1} b_k;$$

Assume P_i is true : $\displaystyle\sum_{k=1}^{i} a_k - b_k = \sum_{k=1}^{i} a_k - \sum_{k=1}^{i} b_k$

$$\sum_{k=1}^{i+1} (a_k - b_k) = \sum_{k=1}^{i} (a_k - b_k) + (a_{i+1} - b_{i+1})$$

$$= \sum_{k=1}^{i} a_k - \sum_{k=1}^{i} b_k + (a_{i+1} - b_{i+1}) \quad \text{(By hypothesis)}$$

$$= \left(\sum_{k=1}^{i} a_k + a_{i+1} \right) - \left(\sum_{k=1}^{i} b_k + b_{i+1} \right)$$

$$= \sum_{k=1}^{i+1} a_k - \sum_{k=1}^{i+1} b_k$$

Thus, P_{i+1} is true and the proof by math induction is complete.

(iii) Proof of (iii) by mathematical induction.

P_1 is true since $\displaystyle\sum_{k=1}^{1} ca_k = ca_1 = c\,(a_1) = c\sum_{k=1}^{1} a_k;$

Assume P_i is true : $\displaystyle\sum_{k=1}^{i} ca_k = c\sum_{k=1}^{i} a_k$

$$\sum_{k=1}^{i+1} ca_k = \sum_{k=1}^{i} ca_k + (ca_{i+1})$$

$$= c\sum_{k=1}^{i} a_k + (ca_{i+1}) \quad \text{(By hypothesis)}$$

$$= c\left(\sum_{k=1}^{i} a_k + a_{i+1}\right)$$

$$= c\sum_{k=1}^{i+1} a_k$$

Thus, P_{i+1} is true and the proof by math induction is complete.

Calculator Exercises 9.3

1) As k increases, the terms approach 1.

4) $x_2 = 3.5$, $x_3 = 3.178571429$, $x_4 = 3.162319422$,

$x_5 = 3.162277661$, $x_6 = 3.162277660$; or as rationals :

$x_2 = \frac{7}{2}$, $x_3 = \frac{89}{28}$, $x_4 = \frac{15,761}{4,984}$, $x_5 = \frac{496,811,681}{157,105,648}$, $x_6 = \frac{493,643,692,713,044,801}{156,103,842,154,948,576}$

7) (a) $y_n = K$ for all n

(b) The answers were obtained by rounding off to 1 decimal place for
each term and then using that value for the next term.

400.0, 560.0, 425.6, 552.3, 436.8,

547.2, 443.9, 543.5, 448.9, 540.7

The terms appear to be oscillating about 500.

Exercises 9.4

1) $d = 4$; $a_n = 2 + (n - 1)(4) = 4n - 2$; $a_5 = 18$; $a_{10} = 38$

4) $d = 1.5$; $a_n = -6 + (n - 1)(1.5) = 1.5n - 7.5$; $a_5 = 0$; $a_{10} = 7.5$

7) An equivalent sequence is $\ln 3$, $2 \ln 3$, $3 \ln 3$, $4 \ln 3$, ...; $d = \ln 3$;

$\quad a_n = \ln 3 + (n - 1)(\ln 3) = n \ln 3$; $a_5 = 5 \ln 3$ or $\ln 243$;

$\quad a_{10} = 10 \ln 3$ or $\ln 59,049$

10) $d = 1 - \sqrt{2}$; $a_n = 2 + \sqrt{2} + (n - 1)(1 - \sqrt{2}) =$

$\quad (1 - \sqrt{2})n + (1 + 2\sqrt{2})$; $a_{11} = 12 - 9\sqrt{2}$

In Exercises 13 & 16, $S_n = \frac{n}{2}[2a_1 + (n - 1)(d)]$ is used to find the sum of the arithmetic sequence.

13) $S_{30} = \frac{30}{2}[2(40) + (29)(-3)] = -105$

16) $a_7 = a_1 + 6d \Rightarrow \frac{7}{3} = a_1 + 6(-\frac{2}{3}) \Rightarrow a_1 = \frac{19}{3}$;

$\quad S_{15} = \frac{15}{2}[2(\frac{19}{3}) + (14)(-\frac{2}{3})] = 25$

In Exercises 17-20, $S_n = \frac{n}{2}(a_1 + a_n)$ is used to find the sum.

19) Using $S_n = \frac{n}{2}(a_1 + a_n)$, we have $S_{18} = \frac{18}{2}(\frac{15}{2} + 16) = \frac{423}{2}$.

22) The first integer greater than -500 that is divisible by 33 is -495 $(-15 \cdot 33)$. There are 15 negative integers greater than -500 that are divisible by 33. The sum is $S_{15} = \frac{15}{2}(-495 + (-33)) = -3960$.

25) Five arithmetic means $\Rightarrow 6d = 10 - 2 \Rightarrow d = \frac{4}{3}$; The terms are $2, \frac{10}{3}, \frac{14}{3}, 6, \frac{22}{3}, \frac{26}{3}, 10$.

28) Model this problem as an arithmetic sequence with $a_1 = 30$ and $d = 2$. $S_{10} = \frac{10}{2}[2(30) + (10 - 1)(2)] = 390$; The last ten rows have $10(50) = 500$ seats so the total is 890 seats.

31) $n = 5$, $S_5 = 5000$, $d = -100 \Rightarrow 5000 = \frac{5}{2}[2a_1 + 4(-100)] \Rightarrow$ $2000 = 2a_1 - 400 \Rightarrow a_1 = \$1200$

34) Let f be the linear function, $f(n) = a(n) + b$. The difference between the $(n + 1)$st term and the nth term is

$$f(n + 1) - f(n) = (a(n + 1) + b) - (a(n) + b)$$
$$= an + a + b - an - b$$
$$= a, \quad \therefore \text{ successive terms differ by the same real}$$

number and the sequence is arithmetic.

1) $r = \frac{4}{8} = \frac{1}{2}$; $a_n = 8\left(\frac{1}{2}\right)^{n-1} = 2^3(2^{-1})^{n-1} = 2^{4-n}$;

$a_5 = 2^{-1} = \frac{1}{2}$; $a_8 = 2^{-4} = \frac{1}{16}$

4) $r = \frac{-\sqrt{3}}{1} = -\sqrt{3}$; $a_n = 1(-\sqrt{3})^{n-1}$; $a_5 = (-\sqrt{3})^4 = 9$;

$a_8 = (-\sqrt{3})^7 = -27\sqrt{3}$

7) $r = \frac{-6}{2} = -\frac{3}{2}$; $a_n = 4\left(-\frac{3}{2}\right)^{n-1}$; $a_5 = 4\left(-\frac{3}{2}\right)^4 = \frac{81}{4}$;

$a_8 = 4\left(-\frac{3}{2}\right)^7 = -\frac{2187}{32}$

10) $r = \frac{-\frac{x}{3}}{1} = -\frac{x}{3}$; $a_n = 1\left(-\frac{x}{3}\right)^{n-1}$; $a_5 = \frac{x^4}{81}$; $a_8 = -\frac{x^7}{2187}$

13) $r = \frac{6}{4} = \frac{3}{2}$; $a_6 = 4\left(\frac{3}{2}\right)^5 = \frac{243}{8}$

16) $a_4 = 4$ and $a_7 = 12 \Rightarrow r^3 = \frac{12}{4} = 3 \Rightarrow r = \sqrt[3]{3}$;

$a_{10} = a_7 r^3 = 12(3) = 36$

19) $\displaystyle\sum_{k=0}^{9} \left(-\frac{1}{2}\right)^{k+1} = \sum_{k=1}^{10} \left(-\frac{1}{2}\right)^k = -\frac{1}{2} \cdot \frac{1 - \left(-\frac{1}{2}\right)^{10}}{1 - \left(-\frac{1}{2}\right)} =$

$-\frac{1}{2} \cdot \frac{1023/1024}{3/2} = -\frac{1023}{3072}$

22) Let $a_1 = 20{,}000$ and $r = 1 - \frac{1}{4} = \frac{3}{4}$. {Since the value at the end of the year is 75% of its value at the beginning of the year.}

$a_7 = 20{,}000\left(\frac{3}{4}\right)^6 \approx \$3{,}559.57$

Exercises 9.5

In the following Exercises, let S_∞ denote $\sum\limits_{k=1}^{\infty}$ and equal $\frac{a_1}{1-r}$.

25) $a_1 = 1$, $r = -\frac{1}{2}$, $S_\infty = \frac{1}{1+\frac{1}{2}} = \frac{2}{3}$

28) $a_1 = 1$, $r = -0.1$, $S_\infty = \frac{1}{1+0.1} = \frac{10}{11}$

31) $0.\overline{23} = 0.23 + 0.0023 + 0.000023 + \cdots$;

$a_1 = 0.23$, $r = 0.01$, $S_\infty = \frac{0.23}{1-0.01} = \frac{23}{99}$

34) $10.\overline{55} = 10 + 0.5 + 0.05 + 0.005 + \cdots$;

$a_1 = 0.5$, $r = 0.1$, $S_\infty = \frac{0.5}{1-0.1} = \frac{5}{9}$; $10.\overline{55} = 10 + \frac{5}{9} = \frac{95}{9}$

37) $1.\overline{6124} = 1 + 0.6124 + 0.00006124 + \cdots$;

$a_1 = 0.6124$, $r = 0.0001$, $S_\infty = \frac{0.6124}{1-0.0001} = \frac{6124}{9999}$;

$1.\overline{6124} = 1 + \frac{6124}{9999} = \frac{16,123}{9999}$

40) $a_1 = 24$ and $r = \frac{5}{6}$ $\Rightarrow$ $S_\infty = \frac{24}{1-5/6} = 144$;

Total distance traveled is 144 cm.

43) (a) A half-life of 2 hours means there will be $(\frac{1}{2})(\frac{1}{2}D) = \frac{1}{4}D$ after 4 hours. The amount remaining after n doses (not hours) for a given dose is $a_n = D(\frac{1}{4})^{n-1}$. Since D mg are administered every 4 hours, the amount of the drug in the bloodstream after n doses is $\sum\limits_{k=1}^{n} a_n = \sum\limits_{k=1}^{n} D(\frac{1}{4})^{n-1} = D + \frac{1}{4}D + \cdots + (\frac{1}{4})^{n-1}D$. Since $r = \frac{1}{4} < 1$, S_n may be approximated by $S_\infty = \frac{a_1}{1-r}$ for large n.

$S_\infty = \frac{D}{1-\frac{1}{4}} = \frac{4}{3}D$.

(b) $\frac{4}{3}D = 500 \Rightarrow D = 375$ mg

1) P_1 is true since $3 (1) - 1 = \dfrac{1 (3 (1) + 1)}{2} = 2$; Assume P_k is true :

$$2 + 5 + 8 + \cdots + 3k - 1 = \frac{k (3k + 1)}{2}$$

$$2 + 5 + 8 + \cdots + 3k - 1 + 3k + 2 = \frac{k (3k + 1)}{2} + 3k + 2$$

$$= \frac{3k^2 + k + 6k + 4}{2}$$

$$= \frac{3k^2 + 7k + 4}{2}$$

$$= \frac{(k + 1) (3k + 4)}{2}$$

Thus, P_{k+1} is true and the proof by math induction is complete.

4) P_1 is true since $(1) ((1) + 1) = \dfrac{(1) ((1) + 1) ((1) + 2)}{3} = 2$;

Assume P_k is true :

$$1 \cdot 2 + 2 \cdot 3 + 3 \cdot 4 + \cdots + k (k + 1) = \frac{k (k + 1) (k + 2)}{3}$$

$$1 \cdot 2 + 2 \cdot 3 + 3 \cdot 4 + \cdots + k (k + 1) + (k + 1) (k + 2) = \frac{k (k + 1) (k + 2)}{3}$$

$$+ (k + 1) (k + 2)$$

$$= (k + 1) (k + 2) \left[\frac{k}{3} + \frac{3}{3} \right]$$

$$= \frac{(k + 1) (k + 2) (k + 3)}{3}$$

Thus, P_{k+1} is true and the proof by math induction is complete.

7) $(x^2 - 3y)^6 = \displaystyle\sum_{k=0}^{6} \binom{6}{k} (x^2)^{6-k} (-3y)^k = \binom{6}{0} (x^2)^6 (-3y)^0 +$

$\binom{6}{1} (x^2)^5 (-3y)^1 + \binom{6}{2} (x^2)^4 (-3y)^2 + \binom{6}{3} (x^2)^3 (-3y)^3 + \binom{6}{4} (x^2)^2 (-3y)^4$

$+ \binom{6}{5} (x^2)^1 (-3y)^5 + \binom{6}{6} (x^2)^0 (-3y)^6$

$= x^{12} - 18x^{10}y + 135x^8y^2 - 540x^6y^3 + 1215x^4y^4 - 1458x^2y^5 + 729y^6$

Exercises 9.6

10) Consider only the variable c in the expansion :

$$(c^3)^{10-k+1} \, (c^{-2})^{k-1} = c^0 \implies c^{35-5k} = c^0 \implies k = 7; \quad a = 2c^3,$$

$$b = 5c^{-2}, \, n = 10; \quad \text{The 7th term is } \binom{n}{k-1} (a)^{n-k+1} \, (b)^{k-1} =$$

$$\binom{10}{6} (2c^3)^4 \, (5c^{-2})^6 = 210 \, (2^4) \, (5^6) = 52{,}500{,}000$$

13) $a_1 = 2$, $a_2 = \frac{1}{2}$, $a_3 = \frac{5}{4}$, $a_4 = \frac{7}{8}$; $\quad a_7 = \frac{65}{64}$

16) $a_1 = 2$, $\qquad a_2 = a_1! = 2! = 2$, $\qquad a_3 = a_2! = 2! = 2$,

$\quad a_4 = a_3! = 2! = 2$, $\qquad a_5 = a_4! = 2! = 2$

19) $5 + 8 + 13 + 20 + 29 = 75$

22) $(-8) + (-6) + (-2) + 6 = -10$

25) $\displaystyle\sum_{k=1}^{5} (-1)^{k+1} \, (105 - 5k) \quad \text{or} \quad \sum_{k=0}^{4} (-1)^k \, (100 - 5k)$

28) $a_4 = a_1 + 3d \implies 9 = a_1 - 15 \implies a_1 = 24$;

$\quad S_8 = \frac{8}{2} [\, 2 \, (24) + (8 - 1) \, (-5) \,] = 52$

31) $r = \dfrac{1/4}{1/8} = 2$; $\quad a_n = \frac{1}{8} (2)^{n-1} = 2^{n-4} \implies a_{10} = 2^6 = 64$

34) $a_8 = a_1 r^7 \implies 100 = a_1 \, (-\frac{3}{2})^7 \implies a_1 = -\frac{12{,}800}{2{,}187}$

37) $\displaystyle\sum_{k=1}^{10} (2^k - \frac{1}{2}) = \sum_{k=1}^{10} 2^k + \sum_{k=1}^{10} (-\frac{1}{2}) = 2 \cdot \frac{1 - 2^{10}}{1 - 2} + 10 \, (-\frac{1}{2})$

$$= 2046 - 5 = 2041$$

40) $6.\overline{274} = 6 + 0.274 + 0.000274 + \cdots$; $\quad a_1 = 0.274$, $r = 0.001$,

$\quad S_\infty = \dfrac{0.274}{1 - 0.001} = \dfrac{274}{999}$; $\quad 6.\overline{274} = 6 + \dfrac{274}{999} = \dfrac{6268}{999}$

Let V, F, and ℓ denote the vertex, focus, and directrix in the following problems.

1) $(x - 0)^2 = -12 (y - 0) \implies 4p = -12 \implies p = -3;$

   V $(0, 0)$; F $(0, -3)$; $\ell : y = 3$; See Figure 1.

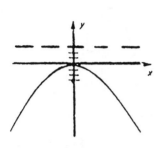
Figure 1

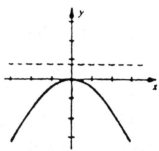
Figure 4

4) $(x - 0)^2 = -3 (y - 0) \implies 4p = -3 \implies p = -\frac{3}{4};$

   V $(0, 0)$; F $(0, -\frac{3}{4})$; $\ell : y = \frac{3}{4}$; See Figure 4.

7) $y^2 - 12 = 12x \implies (y - 0)^2 = 12 (x + 1) \implies 4p = 12 \implies p = 3;$

   V $(-1, 0)$; F $(2, 0)$; $\ell : x = -4$; See Figure 7.

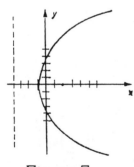
Figure 7

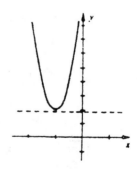
Figure 10

10) $y = 8x^2 + 16x + 10 = 8 (x^2 + 2x + 1) + 10 - 8 \implies$

   $(y - 2) = 8 (x + 1)^2 \implies (x + 1)^2 = \frac{1}{8} (y - 2) \implies 4p = \frac{1}{8} \implies p = \frac{1}{32};$

   V $(-1, 2)$; F $(-1, \frac{65}{32})$; $\ell : y = \frac{63}{32}$; See Figure 10.

Exercises 10.2

13) $-y = 4x^2 + 40x + 106 = 4(x^2 + 10x + 25) + 106 - 100 \Rightarrow$

$-y - 6 = 4(x + 5)^2 \Rightarrow (x + 5)^2 = -\frac{1}{4}(y + 6) \Rightarrow 4p = -\frac{1}{4} \Rightarrow$

$p = -\frac{1}{16}$; $V(-5, -6)$; $F(-5, -\frac{97}{16})$; $l: y = -\frac{95}{16}$; See Figure 13.

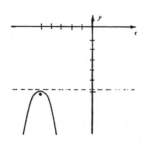

| Figure 13 | Figure 16 |

16) $4x^2 + 4x + 4y + 1 = 0 \Rightarrow -4y = 4(x^2 + x + \frac{1}{4}) + 1 - 1 \Rightarrow$

$(x + \frac{1}{2})^2 = -1(y - 0) \Rightarrow 4p = -1 \Rightarrow p = -\frac{1}{4}$;

$V(-\frac{1}{2}, 0)$; $F(-\frac{1}{2}, -\frac{1}{4})$; $l: y = \frac{1}{4}$; See Figure 16.

19) $F(6, 4)$ and $l: y = -2$; The distance from F to l is 6 units and the focus is above the directrix. This implies that the distance from the directrix to the vertex, p, is 3, and that the vertex is at $(6, 1)$.

$(x - 6)^2 = 4p(y - 1) \Rightarrow (x - 6)^2 = 12(y - 1)$

22) The vertex at $(-3, 5)$ and axis parallel to the x-axis imply that the equation is of the form $(y - 5)^2 = 4p(x + 3)$. Substituting $x = 5$ and $y = 9$ into that equation yields $16 = 4p \cdot 8 \Rightarrow p = \frac{1}{2}$. Thus, the equation is $(y - 5)^2 = 2(x + 3)$.

Exercises 10.3

In this section, V denotes the vertices, F the foci, and M the endpoints of the minor axis.

1) $c^2 = 9 - 4 \Rightarrow c = \pm\sqrt{5}$; $V(\pm 3, 0)$; $F(\pm\sqrt{5}, 0)$; $M(0, \pm 2)$;

See Figure 1.

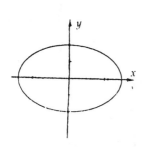

Figure 1

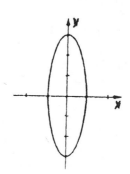

Figure 4

4) $y^2 + 9x^2 = 9 \Rightarrow \dfrac{x^2}{1} + \dfrac{y^2}{9} = 1;\ c^2 = 9 - 1 \Rightarrow c = \pm 2\sqrt{2}$;

   $V(0, \pm 3);\ F(0, \pm 2\sqrt{2});\ M(\pm 1, 0);$ See Figure 4.

7) $4x^2 + 25y^2 = 1 \Rightarrow \dfrac{x^2}{\frac{1}{4}} + \dfrac{y^2}{\frac{1}{25}} = 1;\ c^2 = \frac{1}{4} - \frac{1}{25} = \frac{21}{100} \Rightarrow c =$

   $\pm \sqrt{21}/10;\ V(\pm\frac{1}{2}, 0);\ F(\pm \sqrt{21}/10, 0);\ M(0, \pm\frac{1}{5});$ See Figure 7.

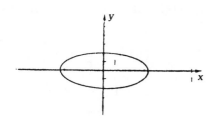

Figure 7

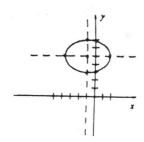

Figure 10

10) $x^2 + 2y^2 + 2x - 20y + 43 = 0 \Rightarrow$

   $(x^2 + 2x + 1) + 2(y^2 - 10y + 25) = -43 + 1 + 50 \Rightarrow$

   $\dfrac{(x + 1)^2}{8} + \dfrac{(y - 5)^2}{4} = 1;\ c^2 = 8 - 4 \Rightarrow c = \pm 2;$ Center $(-1, 5);$

   $V(-1 \pm 2\sqrt{2}, 5);\ F(1, 5);\ F'(-3, 5);\ M(-1, 7);\ M'(-1, 3);$

   See Figure 10.

Exercises 10.3

13) $25x^2 + 4y^2 - 250x - 16y + 541 = 0$ ⟹

   $25(x^2 - 10x + 25) + 4(y^2 - 4y + 4) = -541 + 625 + 16$ ⟹

   $\dfrac{(x-5)^2}{4} + \dfrac{(y-2)^2}{25} = 1$; $c^2 = 25 - 4$ ⟹ $c = \pm\sqrt{21}$; $C(5, 2)$;

   $V(5, 7)$; $V'(5, -3)$; $F(5, 2 \pm \sqrt{21})$; $M(7, 2)$; $M'(3, 2)$

   See Figure 13.

16) $b^2 = 7^2 - 2^2 = 45$; An

   equation of the ellipse is

   $\dfrac{x^2}{45} + \dfrac{y^2}{49} = 1$.

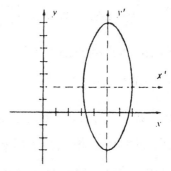

Figure 13

19) With the vertices at $(0, \pm 6)$, an equation of the ellipse is

   $\dfrac{x^2}{b^2} + \dfrac{y^2}{36} = 1$. Substituting $x = 3$ and $y = 2$ and solving for b^2 yields

   $\dfrac{9}{b^2} + \dfrac{4}{36} = 1$ ⟹ $\dfrac{9}{b^2} = \dfrac{8}{9}$ ⟹ $b^2 = \dfrac{81}{8}$. An equation is $\dfrac{8x^2}{81} + \dfrac{y^2}{36} = 1$.

22) E1 - E2 ⟹ $3y^2 = 24$ ⟹ $y^2 = 8$; $x^2 = 4$; The 4 intersection

   points are $(\pm 2, \pm 2\sqrt{2})$. See Figure 22.

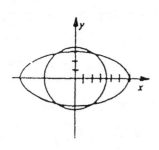

Figure 22

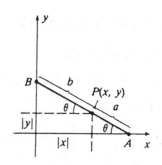

Figure 25

25) To picture the movement, let A approach O while B moves in the positive y direction. The maximum y value of P occurs when A = O and y = a. A continues in the negative x direction and now B will drop from $(0, a + b)$ to $(0, -(a + b))$. We wish to show that the path traced by P (x, y) is that of an ellipse. Let $\theta = \angle BAO$. Now

$$\cos\theta = \frac{|x|}{b} \text{ and } \sin\theta = \frac{|y|}{a} \Rightarrow \cos^2\theta + \sin^2\theta = \frac{x^2}{b^2} + \frac{y^2}{a^2} = 1,$$

which is the equation of an ellipse if $a \neq b$. {A circle if $a = b$} See Figure 25.

28) Note that the y-axis is misplaced in the figure in the text. The circle has the coordinates $(x, y) = (x, \pm\sqrt{4 - x^2})$. The midpoints have the coordinates $(x, y') = (x, \pm\frac{1}{2}\sqrt{4 - x^2})$. Now

$$x^2 + (y')^2 = x^2 + (\tfrac{1}{2}y)^2 \quad \text{(since } y' = \tfrac{1}{2}y)$$
$$= \tfrac{3}{4}x^2 + (\tfrac{1}{4}x^2 + \tfrac{1}{4}y^2)$$
$$= \tfrac{3}{4}x^2 + 1 \quad \text{(since } x^2 + y^2 = 4)$$

This implies $\tfrac{1}{4}x^2 + (y')^2 = 1$ or equivalently, $x^2 + 4(y')^2 = 4$ which is an equation of an ellipse.

In this section, V denotes the vertices, F the foci, and W the endpoints of the conjugate axis.

1) $c^2 = 9 + 4 \Rightarrow c = \pm\sqrt{13}$; V $(\pm 3, 0)$; W $(0, \pm 2)$; F $(\pm\sqrt{13}, 0)$; $y = \pm\tfrac{2}{3}x$; See Figure 1.

4) $c^2 = 49 + 16 \Rightarrow c = \pm\sqrt{65}$; V $(\pm 7, 0)$; W $(0, \pm 4)$; F $(\pm\sqrt{65}, 0)$; $y = \pm\tfrac{4}{7}x$; See Figure 4.

Exercises 10.4

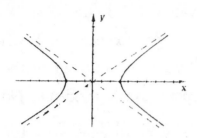

Figure 1

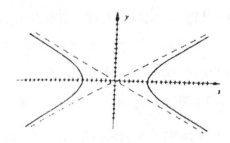

Figure 4

7) $c^2 = 1 + 1 \Rightarrow c = \pm\sqrt{2}$; $V(\pm 1, 0)$; $W(0, \pm 1)$; $F(\pm\sqrt{2}, 0)$;

$y = \pm x$; See Figure 7.

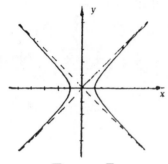

Figure 7

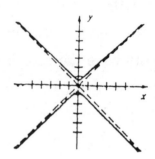

Figure 10

10) $4y^2 - 4x^2 = 1 \Rightarrow \frac{y^2}{\frac{1}{4}} - \frac{x^2}{\frac{1}{4}} = 1$; $c^2 = \frac{1}{4} + \frac{1}{4} \Rightarrow c = \pm\sqrt{2}/2$;

$V(0, \pm\frac{1}{2})$; $W(\pm\frac{1}{2}, 0)$; $F(0, \pm\sqrt{2}/2)$; $y = \pm x$; See Figure 10.

13) $25x^2 - 16y^2 + 250x + 32y + 109 = 0 \Rightarrow$

$25(x^2 + 10x + 25) - 16(y^2 - 2y + 1) = -109 + 625 - 16 = 500$

$\Rightarrow \frac{(x + 5)^2}{20} - \frac{(y - 1)^2}{\frac{125}{4}} = 1$; $c^2 = 20 + \frac{125}{4} \Rightarrow c = \pm\sqrt{205}/2$;

$C(-5, 1)$; $V(-5 \pm 2\sqrt{5}, 1)$; $W(-5, 1 \pm 5\sqrt{5}/2)$;

$F(-5 \pm \sqrt{205}/2, 1)$; slope of asymptotes $= \pm\dfrac{\sqrt{125/4}}{\sqrt{20}} = \pm\dfrac{5}{4}$;

(continued on next page)

198

Since the center is (-5, 1), the asymptote equations are
$(y - 1) = \pm \frac{5}{4}(x + 5)$. See Figure 13.

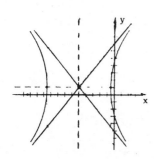

Figure 13

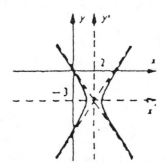

Figure 16

16) $25x^2 - 9y^2 - 100x - 54y + 10 = 0 \Rightarrow$

$25(x^2 - 4x + 4) - 9(y^2 + 6y + 9) = -10 + 100 - 81 = 9 \Rightarrow$

$\frac{(x - 2)^2}{\frac{9}{25}} - \frac{(y + 3)^2}{1} = 1$; $c^2 = \frac{9}{25} + 1 \Rightarrow c = \pm \sqrt{34}/5$; $C(2, -3)$;

$V(2 \pm \frac{3}{5}, -3)$; $W(2, -3 \pm 1)$; $F(2 \pm \sqrt{34}/5, -3)$; slope of

asymptotes $= \pm \dfrac{\sqrt{1}}{\sqrt{9/25}} = \pm \dfrac{5}{3}$; Since the center is $(2, -3)$, the

asymptote equations are $(y + 3) = \pm \frac{5}{3}(x - 2)$; See Figure 16.

19) $F(0, \pm 4)$ and $V(0, \pm 1) \Rightarrow W(\pm \sqrt{15}, 0)$; $\{4^2 - 1^2 = (\sqrt{15})^2\}$

An equation is $\frac{y^2}{1} - \frac{x^2}{15} = 1$.

22) $F(0, \pm 3)$ and $V(0, \pm 2) \Rightarrow W(\pm \sqrt{5}, 0)$; $\{3^2 - 2^2 = (\sqrt{5})^2\}$

An equation is $\frac{y^2}{4} - \frac{x^2}{5} = 1$.

25) Asymptote equations of $y = \pm 2x$ and $V(\pm 3, 0) \Rightarrow W(0, \pm 6)$;

An equation is $\frac{x^2}{9} - \frac{y^2}{36} = 1$.

Exercises 10.4

28) Adding the two equations yields $x^2 - 3x = 4 \Rightarrow x^2 - 3x - 4 = 0$
$\Rightarrow (x - 4)(x + 1) = 0 \Rightarrow x = 4, -1$. Substituting these values in
the second equation, we find that for $x = 4$, $y = \pm 2\sqrt{3}$, and for
$x = -1$, there are no real solutions for y. The two points of
intersection are $(4, \pm 2\sqrt{3})$. See Figure 28.

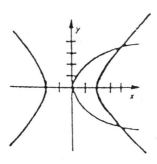

Figure 28

31) The path is a hyperbola with $V(\pm 3, 0)$ and $W(0, \pm \frac{3}{2})$. An equation
is $\frac{x^2}{9} - \frac{y^2}{\frac{9}{4}} = 1$ or $x^2 - 4y^2 = 9$. If only the right branch is
considered, then $x = \sqrt{9 + 4y^2}$ is an equation of the path.

Exercises 10.5

The following is a general outline of the solutions for Exercises 1-14 in
this section and 45-46 in the Chapter Review.

(a) $\cot 2\phi = \frac{A - C}{B} \Rightarrow \phi = \frac{1}{2} \cot^{-1}\left(\frac{A - C}{B}\right)$ where the range of $\cot^{-1}$ is
$(0, \Pi)$.

(b) $\cos 2\phi = \frac{\pm (A - C)}{\sqrt{(A - C)^2 + B^2}}$, $\sin \phi = \sqrt{\frac{1 - \cos 2\phi}{2}}$, and

$\cos \phi = \sqrt{\frac{1 + \cos 2\phi}{2}}$; Note that $\cos 2\phi$ will have the same sign as cot 2ϕ since cot $2\phi = \dfrac{\cos 2\phi}{\sin 2\phi}$ and sin 2ϕ is positive. Also, note that sin ϕ and cos ϕ will always be positive since ϕ is $\frac{1}{2}$ of an angle between 0 and Π. { the range of cot^{-1} }

(c) The Rotation of Axes Formulas are

$$\begin{cases} x = x' \cos \phi - y' \sin \phi \\ y = x' \sin \phi + y' \cos \phi \end{cases}$$

(d) The Rotation of Axes Formulas are substituted into the original equation to obtain an equation in x' and y'. This equation is then simplified into a standard form.

(e) Pertinent information about the graph on the $x'y'$-plane is listed.

Parabola : (V)ertex and (F)ocus

Ellipse : (C)enter, (V)ertices, and endpoints of the minor axis (M)

Hyperbola : (C)enter, (V)ertices, endpoints of the conjugate axis (W), and the equations of the asymptotes

(f) The information found in part (e) is found for the xy-plane by substituting the values from the $x'y'$-plane into the Rotation of Axes Formulas. The work for the asymptotes is shown.

Rather than presenting the normal pattern of problems, we include Exercises 1, 2, 4, 7, 9, and 11. This is a balanced distribution of 2 ellipses, 2 hyperbolas, and 2 parabolas.

1) (a) cot $2\phi = \frac{7}{24}$; $\phi \approx 36.87°$; See Figure 1.

(b) cos $2\phi = \frac{7}{25}$; sin $\phi = \frac{3}{5}$; cos $\phi = \frac{4}{5}$;

(c) $$\begin{cases} x = \frac{4}{5} x' - \frac{3}{5} y' = \frac{1}{5} (4x' - 3y') \\ y = \frac{3}{5} x' + \frac{4}{5} y' = \frac{1}{5} (3x' + 4y') \end{cases}$$

Exercises 10.5

(d) $\frac{125}{25}(x')^2 + \frac{2000}{25}(y')^2 = 80 \implies \frac{(x')^2}{16} + \frac{(y')^2}{1} = 1$; ellipse

(e) $C'(0, 0)$; $V'(\pm 4, 0)$; $M'(0, \pm 1)$

(f) $C(0, 0)$; $V(\pm \frac{16}{5}, \pm \frac{9}{5})$; $M(\mp \frac{3}{5}, \pm \frac{4}{5})$

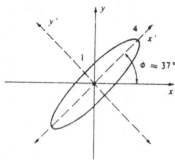

Figure 1

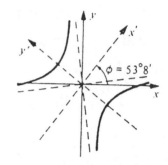

Figure 2

2) (a) $\cot 2\phi = -\frac{7}{24}$; $\phi \approx 53.13°$; See Figure 2.

(b) $\cos 2\phi = -\frac{7}{25}$; $\sin \phi = \frac{4}{5}$; $\cos \phi = \frac{3}{5}$

(c) $\begin{cases} x = \frac{3}{5}x' - \frac{4}{5}y' = \frac{1}{5}(3x' - 4y') \\ y = \frac{4}{5}x' + \frac{3}{5}y' = \frac{1}{5}(4x' + 3y') \end{cases}$

(d) $-\frac{625}{25}(x')^2 + \frac{625}{25}(y')^2 = 225 \implies \frac{(y')^2}{9} - \frac{(x')^2}{9} = 1$; hyperbola

(e) $C'(0, 0)$; $V'(0, \pm 3)$; $W'(\pm 3, 0)$; $y' = \pm x'$

(f) $C(0, 0)$; $V(\mp \frac{12}{5}, \pm \frac{9}{5})$; $W(\pm \frac{9}{5}, \pm \frac{12}{5})$; For the asymptotes :

$x = \frac{3}{5}x' - \frac{4}{5}y' = \frac{3}{5}x' - \frac{4}{5}(\pm x') = -\frac{1}{5}x', \frac{7}{5}x' \implies x' = -5x, \frac{5}{7}x$;

$y = \frac{4}{5}x' + \frac{3}{5}y' = \frac{4}{5}x' + \frac{3}{5}(\pm x') = \frac{7}{5}x', \frac{1}{5}x' = \frac{7}{5}(-5x), \frac{1}{5}(\frac{5}{7}x)$

$= -7x, \frac{1}{7}x$

4) (a) $\cot 2\phi = 0$; $\phi = 45°$; See Figure 4.

(b) $\cos 2\phi = 0$; $\sin \phi = \sqrt{2}/2$; $\cos \phi = \sqrt{2}/2$

(c) $\begin{cases} x = (\sqrt{2}/2) x' - (\sqrt{2}/2) y' = (\sqrt{2}/2)(x' - y') \\ y = (\sqrt{2}/2) x' + (\sqrt{2}/2) y' = (\sqrt{2}/2)(x' + y') \end{cases}$

(d) $\frac{1}{2}(x')^2 + \frac{3}{2}(y')^2 = 3 \implies \frac{(x')^2}{6} + \frac{(y')^2}{2} = 1$; ellipse

(e) $C'(0, 0)$; $V'(\pm\sqrt{6}, 0)$; $M'(0, \pm\sqrt{2})$

(f) $C(0, 0)$; $V(\pm\sqrt{3}, \pm\sqrt{3})$; $M(\mp 1, \pm 1)$

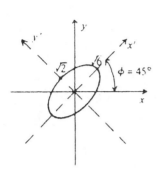

Figure 4

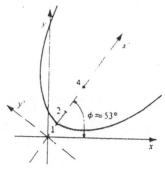

Figure 7

7) (a) $\cot 2\phi = -\frac{7}{24}$; $\phi \approx 53.13°$; See Figure 7.

(b) $\cos 2\phi = -\frac{7}{25}$; $\sin\phi = \frac{4}{5}$; $\cos\phi = \frac{3}{5}$

(c) $\begin{cases} x = \frac{3}{5}x' - \frac{4}{5}y' = \frac{1}{5}(3x' - 4y') \\ y = \frac{4}{5}x' + \frac{3}{5}y' = \frac{1}{5}(4x' + 3y') \end{cases}$

(d) $\frac{625}{25}(y')^2 - 100x' + 100 = 0 \implies (y')^2 = 4(x' - 1)$;

parabola with $p = 1$

(e) $V'(1, 0)$; $F'(2, 0)$

(f) $V(\frac{3}{5}, \frac{4}{5})$; $F(\frac{6}{5}, \frac{8}{5})$

9) (a) $\cot 2\phi = 1/\sqrt{3}$; $\phi = 30°$; See Figure 9.

(b) $\cos 2\phi = \frac{1}{2}$; $\sin\phi = \frac{1}{2}$; $\cos\phi = \sqrt{3}/2$

(c) $\begin{cases} x = (\sqrt{3}/2)x' - \frac{1}{2}y' = \frac{1}{2}(\sqrt{3}x' - y') \\ y = \frac{1}{2}x' + (\sqrt{3}/2)y' = \frac{1}{2}(x' + \sqrt{3}y') \end{cases}$

(d) $\frac{32}{4} (x')^2 - \frac{16}{4} (y')^2 - 16y' - 12 = 0 \Rightarrow$

$2 (x')^2 - (y')^2 - 4y' - 3 = 0 \Rightarrow$

$2 (x')^2 - ((y')^2 + 4y' + 4) - 3 + 4 = 0 \Rightarrow$

$\dfrac{(y' + 2)^2}{1} - \dfrac{(x')^2}{\frac{1}{2}} = 1$; hyperbola

(e) $C'(0, -2)$; $V'(0, -2 \pm 1)$; $W'(\pm (\sqrt{2}/2), -2)$; $y' = \pm \sqrt{2} x'$

(f) $C (1, -\sqrt{3})$; $V (1 \mp \frac{1}{2}, -\sqrt{3} \pm (\sqrt{3}/2))$;

$W (\pm (\sqrt{6}/4) + 1, \pm (\sqrt{2}/4) - \sqrt{3})$; Asymptotes :

$x = (\sqrt{3}/2) x' - \frac{1}{2} y' = (\sqrt{3}/2) x' - \frac{1}{2} (\pm \sqrt{2} x') = \left(\dfrac{\sqrt{3} \mp \sqrt{2}}{2} \right) x'$

$\Rightarrow x' = 2 (\sqrt{3} \pm \sqrt{2}) x;$

$y = \frac{1}{2} x' + (\sqrt{3}/2) y' = \frac{1}{2} x' + (\sqrt{3}/2) (\pm \sqrt{2} x')$

$= \left(\dfrac{1 \pm \sqrt{6}}{2} \right) x' = \left(\dfrac{1 \pm \sqrt{6}}{2} \right) (2 (\sqrt{3} \pm \sqrt{2})) x = (3 \sqrt{3} \pm 4 \sqrt{2}) x$

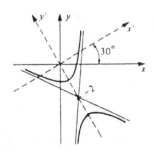

Figure 9   Figure 11

11) (a) $\cot 2\phi = -\frac{3}{4}$; $\phi \approx 63.43°$; See Figure 11.

(b) $\cos 2\phi = -\frac{3}{5}$; $\sin \phi = 2/\sqrt{5}$; $\cos \phi = 1/\sqrt{5}$

(c) $\begin{cases} x = (1/\sqrt{5}) x' - (2/\sqrt{5}) y' = (1/\sqrt{5}) (x' - 2y') \\ y = (2/\sqrt{5}) x' + (1/\sqrt{5}) y' = (1/\sqrt{5}) (2x' + y') \end{cases}$

(d) $\frac{25}{5} (x')^2 - 30x' - 30y' + 45 = 0 \Rightarrow (x' - 3)^2 = 6y';$

parabola with p = $\frac{3}{2}$

(e) V'(3, 0); F'(3, $\frac{3}{2}$)

(f) V (3/$\sqrt{5}$, 6/$\sqrt{5}$); F (0, 3 $\sqrt{5}$/2)

16)

Exercise #	$B^2 - 4AC$	Type of graph
1	−1600	Ellipse
2	2500	Hyperbola
3	256	Hyperbola
4	−3	Ellipse
5	−36	Ellipse
6	256	Hyperbola
7	0	Parabola
8	0	Parabola
9	128	Hyperbola
10	−3600	Ellipse
11	0	Parabola
12	$400 + 240\sqrt{2}$	Hyperbola
13	−2704	Ellipse
14	0	Parabola

For the following exercises, the substitutions y = r sin θ, x = r cos θ, $r^2 = x^2 + y^2$, and tan θ = $\frac{y}{x}$ are used without mention. I have found it helpful to find the "pole" values { when the graph intersects the pole } to determine which values of θ should be used in the construction of an r-θ chart. The numbers listed on each line of the r-θ chart correspond to the numbers labeled on the figures.

1) r = 5 $\Rightarrow$ $r^2 = 25$ $\Rightarrow$ $x^2 + y^2 = 25$; {circle} See Figure 1.

2) θ = Π/4 and r $\in$ (−∞, ∞); The line is y = (tan θ) x = x. See Figure 2.

Exercises 10.6

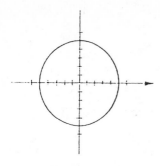

Figure 1

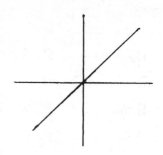

Figure 2

7) $r = 4(1 - \sin\theta)$ is a cardiod. $0 = 4(1 - \sin\theta) \Rightarrow \sin\theta = 1 \Rightarrow$
$\theta = \Pi/2$

	Range of θ			Range of r	
1)	0	→	$\Pi/2$	4 →	0
2)	$\Pi/2$	→	Π	0 →	4
3)	Π	→	$3\Pi/2$	4 →	8
4)	$3\Pi/2$	→	2Π	8 →	4

See Figure 7.

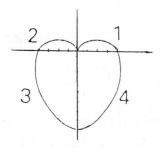

Figure 7

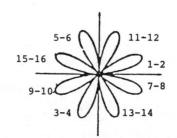

Figure 10

10) $r = 2\sin 4\theta$ is an 8-leafed rose. $0 = 2\sin 4\theta \Rightarrow \sin 4\theta = 0 \Rightarrow$
$4\theta = \Pi n \Rightarrow \theta = (\Pi/4) n$. Steps 9 through 16 follow a similar
pattern to steps 1 through 8 and are labeled in the correct order.

	Range of θ		Range of r
1)	0 →	Π/8	0 → 2
2)	Π/8 →	Π/4	2 → 0
3)	Π/4 →	3Π/8	0 → -2
4)	3Π/8 →	Π/2	-2 → 0
5)	Π/2 →	5Π/8	0 → 2
6)	5Π/8 →	3Π/4	2 → 0
7)	3Π/4 →	7Π/8	0 → -2
8)	7Π/8 →	Π	-2 → 0

See Figure 10.

13) $r = 4 \csc \theta \Rightarrow r \sin \theta = 4 \Rightarrow y = 4$; r is undefined at $\theta = \Pi n$
    See Figure 13.

Figure 13   Figure 16

16) $r = 2 \sec \theta$ is equivalent to $x = 2$ and r is undefined at
    $\theta = (\Pi/2) + \Pi n$; $0 = 2 + 2 \sec \theta \Rightarrow \sec \theta = -1 \Rightarrow \theta = \Pi$

	Range of θ		Range of r
1)	0 →	Π/2	4 → +∞
2)	Π/2 →	Π	-∞ → 0
3)	Π →	3Π/2	0 → -∞
4)	3Π/2 →	2Π	+∞ → 4

See Figure 16.

19) r = -6 (1 + cos θ) is a cardiod; 0 = -6 (1 + cos θ) ⟹ cos θ = -1

⟹ θ = Π

	Range of θ			Range of r	
1)	0	→	Π/2	-12 →	-6
2)	Π/2	→	Π	-6 →	0
3)	Π	→	3Π/2	0 →	-6
4)	3Π/2	→	2Π	-6 → -12	

See Figure 19.

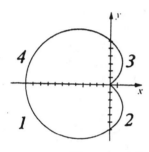

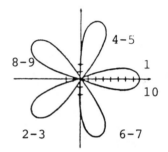

Figure 19   Figure 22

22) r = 8 cos 5θ is a 5-leafed rose. 0 = 8 cos 5θ ⟹ cos 5θ = 0 ⟹
5θ = (Π/2) + Πn ⟹ θ = (Π/10) + (Π/5)n

	Range of θ			Range of r	
1)	0	→	Π/10	8 →	0
2)	Π/10	→	2Π/10	0 →	-8
3)	2Π/10	→	3Π/10	-8 →	0
4)	3Π/10	→	4Π/10	0 →	8
5)	4Π/10	→	5Π/10	8 →	0
6)	5Π/10	→	6Π/10	0 →	-8
7)	6Π/10	→	7Π/10	-8 →	0
8)	7Π/10	→	8Π/10	0 →	8
9)	8Π/10	→	9Π/10	8 →	0
10)	9Π/10	→	Π	0 →	-8

See Figure 22.

25) $x = -3 \Rightarrow r\cos\theta = -3 \Rightarrow r = -3\sec\theta$

28) $x^2 = 8y \Rightarrow r^2\cos^2\theta = 8r\sin\theta \Rightarrow r = 8\tan\theta\sec\theta$ or $r = 0$;

   $r = 0$ is obtained in $8\tan\theta\sec\theta$ if $\theta = 0$.

31) $x^2 - y^2 = 16 \Rightarrow r^2\cos^2\theta - r^2\sin^2\theta = 16 \Rightarrow$

   $r^2(\cos^2\theta - \sin^2\theta) = 16 \Rightarrow r^2\cos 2\theta = 16 \Rightarrow r^2 = 16\sec 2\theta$ or

   $r = \pm 4\sqrt{\sec 2\theta}$

34) $r\sin\theta = -2 \Rightarrow y = -2$; {horizontal line} See Figure 34.

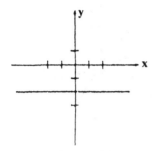

Figure 34   Figure 36

36) $r = 6\cot\theta \Rightarrow r^2 = 36\cot^2\theta \Rightarrow x^2 + y^2 = 36\left(\dfrac{x^2}{y^2}\right) \Rightarrow$

   $y^4 + x^2y^2 - 36x^2 = 0$. Solve this as a quadratic in y^2;

   $y^2 = \dfrac{-x^2 \pm \sqrt{x^4 + 144x^2}}{2}$ {Since $y^2 > 0$, the negative radical solution is extraneous.}

   $y^2 = \dfrac{-x^2 + \sqrt{x^4 + 144x^2}}{2} \cdot \dfrac{-x^2 - \sqrt{x^4 + 144x^2}}{-x^2 - \sqrt{x^4 + 144x^2}}$ {multiplying by the conjugate}

   $= \dfrac{x^4 - (x^4 + 144x^2)}{2(-x^2 - \sqrt{x^4 + 144x^2})}$

   $= \dfrac{-72x^2}{-x^2 - \sqrt{x^4 + 144x^2}}$

   $= \dfrac{-72}{-1 - \sqrt{1 + (144/x^2)}}$ {dividing by x^2}

(continued on next page)

If x gets very large {positively or negatively}, then $144/x^2$ gets close to 0 and the entire fraction gets close to 36. $y^2 = 36 \Rightarrow$ $y = \pm 6$ are horizontal asymptotes for the graph. The following table may be helpful. Note that r is undefined at Πn.

	Range of θ			Range of r		
1)	0	→	$\Pi/2$	$+\infty$	→	0
2)	$\Pi/2$	→	Π	0	→	$-\infty$
3)	Π	→	$3\Pi/2$	$+\infty$	→	0
4)	$3\Pi/2$	→	2Π	0	→	$-\infty$

See Figure 36.

40) $r = 4 \sec \theta \Rightarrow r \cos \theta = 4 \Rightarrow x = 4$; {vertical line}
θ is undefined at $(\Pi/2) + \Pi n$; See Figure 40.

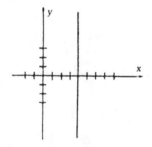

Figure 40

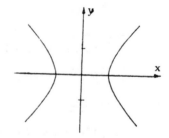

Figure 43

43) $r^2 \cos 2\theta = 1 \Rightarrow r^2 (\cos^2 \theta - \sin^2 \theta) = 1 \Rightarrow$
$r^2 \cos^2 \theta - r^2 \sin^2 \theta = 1 \Rightarrow x^2 - y^2 = 1$; {hyperbola}
See Figure 43.

46) $r (\sin \theta + r \cos^2 \theta) = 1 \Rightarrow r \sin \theta + r^2 \cos^2 \theta = 1 \Rightarrow$
$y + x^2 = 1$ or $y = -x^2 + 1$; {parabola} See Figure 46.

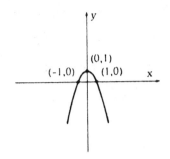

Figure 46   Figure 49

49) Let P_1 (r_1, θ_1) and P_2 (r_2, θ_2) be points in the $r\theta$-plane. Let
$a = r_1$, $b = r_2$, $c = d$ (P_1, P_2), and $\gamma = \theta_2 - \theta_1$. The Law of Cosines
states that $c^2 = a^2 + b^2 - 2ab \cos \gamma$ $(\star)$. By substituting a, b, c, and
γ into $(\star)$, the result in the text immediately follows.
See Figure 49.

For the ellipse the major axis is vertical if the denominator contains
$\sin \theta$, horizontal if the denominator contains $\cos \theta$. For the hyperbola
the transverse axis is vertical if the denominator contains $\sin \theta$,
horizontal if the denominator contains $\cos \theta$. The focus at the pole is
called F and V is the vertex associated with F. d (V, F) denotes the
distance from the vertex to the focus. For the parabola the directrix
is on the right, left, top, or bottom of the focus depending on the
term "+cos", "-cos", "+sin", or "-sin", respectively, appearing in the
denominator.

1) $r = \dfrac{12}{6 + 2 \sin \theta} = \dfrac{2}{1 + \frac{1}{3} \sin \theta}$ $\Rightarrow$ $e = \frac{1}{3}$; ellipse

   V $(\frac{3}{2}, \Pi/2)$ and V'$(3, 3\Pi/2)$; d (V, F) $= \frac{3}{2}$ $\Rightarrow$ F' $= (\frac{3}{2}, 3\Pi/2)$
   See Figure 1.

Exercises 10.7

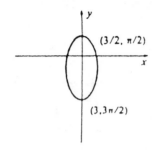

Figure 1

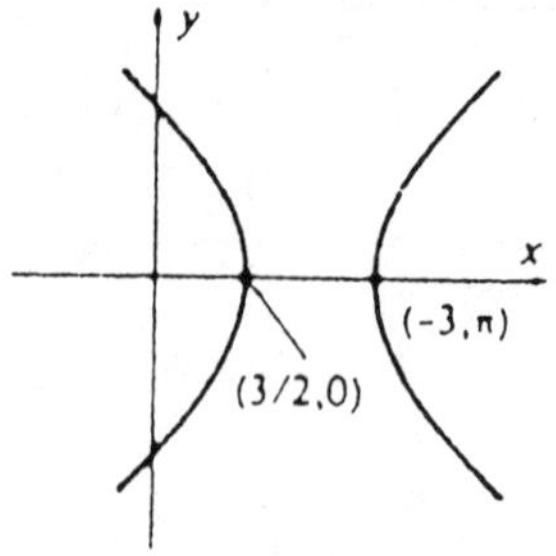

Figure 4

4) $r = \dfrac{12}{2 + 6 \cos \theta} = \dfrac{6}{1 + 3 \cos \theta} \Rightarrow e = 3;$ hyperbola

  $V\left(\frac{3}{2}, 0\right)$ and $V'(-3, \Pi);$ $d(V, F) = \frac{3}{2} \Rightarrow F' = \left(-\frac{3}{2}, \Pi\right);$

  See Figure 4.

7) $r = \dfrac{4}{\cos \theta - 2} = \dfrac{-2}{1 - \frac{1}{2} \cos \theta} \Rightarrow e = \frac{1}{2};$ ellipse

  $V\left(-\frac{4}{3}, \Pi\right)$ and $V'(-4, 0);$ $d(V, F) = \frac{4}{3} \Rightarrow F' = \left(-\frac{8}{3}, 0\right);$

  See Figure 7.

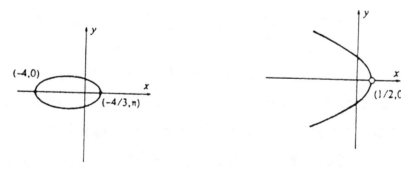

Figure 7

Figure 10

10) $r = \csc \theta \,(\csc \theta - \cot \theta) = \dfrac{1}{\sin \theta}\left(\dfrac{1 - \cos \theta}{\sin \theta}\right) = \dfrac{1 - \cos \theta}{1 - \cos^2 \theta} =$

  $\dfrac{1}{1 + \cos \theta} \Rightarrow e = 1;$ parabola; The vertex is in the $\theta = 0$

  direction. $V\left(\frac{1}{2}, 0\right)$ Since the original equation is undefined when csc θ or cot θ is undefined, $(\frac{1}{2}, 0)$ {the vertex} is excluded from the graph. See Figure 10.

For the following exercises, the substitutions $\cos \theta = \frac{x}{r}$, $\sin \theta = \frac{y}{r}$, and $r^2 = x^2 + y^2$ are made without mention.

13) $r = \dfrac{12}{2 - 6 \cos \theta}$ $\Rightarrow$ $2r - 6x = 12$ $\Rightarrow$ $r = 3x + 6$ $\Rightarrow$

$r^2 = 9x^2 + 36x + 36$ $\Rightarrow$ $8x^2 - y^2 + 36x + 36 = 0$

16) $r = \dfrac{3}{2 - 2 \sin \theta}$ $\Rightarrow$ $2r - 2y = 3$ $\Rightarrow$ $2r = 2y + 3$ $\Rightarrow$

$4r^2 = 4y^2 + 12y + 9$ $\Rightarrow$ $4x^2 - 12y - 9 = 0$

19) $r = \dfrac{6 \csc \theta}{2 \csc \theta + 3} \cdot \dfrac{\sin \theta}{\sin \theta} = \dfrac{6}{2 + 3 \sin \theta}$ $\Rightarrow$ $2r + 3y = 6$ $\Rightarrow$

$2r = 6 - 3y$ $\Rightarrow$ $4r^2 = 36 - 36y + 9y^2$ $\Rightarrow$ $4x^2 - 5y^2 + 36y - 36$

$= 0$; r is undefined when $\theta = 0$ or Π. For the rectangular equation, these points correspond to $y = 0$ {or $r \sin \theta = 0$}. Substituting $y = 0$ into the above rectangular equation yields $4x^2 = 36$ or $x = \pm 3$.

$\therefore$ exclude $(\pm 3, 0)$

22) $x^2 = 1 - 2y = -2 (y - \frac{1}{2})$; $4p = -2$ $\Rightarrow$ $p = -\frac{1}{2}$; Since the vertex is $(0, \frac{1}{2})$, the focus is $(0, 0)$ and the directrix is $y = 1$ or $r = \csc \theta$. The distance from the focus to the directrix, d, is 1. The eccentricity of a parabola is 1. $r = \dfrac{de}{1 + e \sin \theta} = \dfrac{1}{1 + \sin \theta}$ is a polar equation of the parabola. The "$+ \sin \theta$" is chosen since the directrix is on the top of the focus.

25) $8x^2 + 9y^2 + 4x = 4$ $\Rightarrow$ $9x^2 + 9y^2 = x^2 - 4x + 4$ $\Rightarrow$ $9r^2 = (x - 2)^2$

$\Rightarrow$ $\pm 3r = r \cos \theta - 2$ $\Rightarrow$ $\pm 3r - r \cos \theta = -2$ $\Rightarrow$

Exercises 10.7

$$r = \frac{-2}{\pm 3 - \cos\theta} = \frac{\frac{2}{3}}{1 + \frac{1}{3}\cos\theta} \quad \{\text{using the "−"}\}$$

Using the "+" or the "−" leads to the same graph. $e = \frac{1}{3}$ and $de = \frac{2}{3}$
$\Rightarrow d = 2$; The directrix is on the right of the focus at the pole and has the equation $x = 2$ or $r = 2\sec\theta$.

28) $r = 4\csc\theta \Rightarrow d = 4$ and use " $+ \sin\theta$ ";

$$r = \frac{4\,(2/5)}{1 + (2/5)\sin\theta} \cdot \frac{5}{5} = \frac{8}{5 + 2\sin\theta}$$

31) $r = 5\sec\theta \Rightarrow d = 5$ and use " $+ \cos\theta$ "

$$r = \frac{5\,(1)}{1 + 1\cos\theta} = \frac{5}{1 + \cos\theta}$$

34) There are two possible cases. The <u>first</u> is if the vertex $(1, 3\Pi/2)$ is associated with the focus at the pole. The equation would then have the form $r = \dfrac{d\left(\frac{2}{3}\right)}{1 - \frac{2}{3}\sin\theta} = \dfrac{2d}{3 - 2\sin\theta}$ ($\star$) since the directrix would be under the focus. The <u>second</u> case is if the vertex $(1, 3\Pi/2)$ is <u>not</u> associated with the focus at the pole and the other vertex is. The equation would then have the form $r = \dfrac{d\left(\frac{2}{3}\right)}{1 + \frac{2}{3}\sin\theta} = \dfrac{2d}{3 + 2\sin\theta}$ ($\star\star$) since the directrix would be on top of the focus. Substituting $\theta = 3\Pi/2$ and $r = 1$ into ($\star$) and ($\star\star$) yield $d = \frac{5}{2}$ and $\frac{1}{2}$ respectively. The equations are then the two different ellipses :

$$r = \frac{5}{3 - 2\sin\theta} \quad \text{and} \quad r = \frac{1}{3 + 2\sin\theta}.$$

Exercises 10.8

1) $x = t - 2 \Rightarrow t = x + 2$; $y = 2(x + 2) + 3 = 2x + 7$; As t varies from 0 to 5, (x, y) varies from $(-2, 3)$ to $(3, 13)$. See Figure 1.

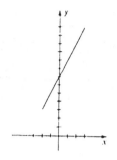

Figure 1

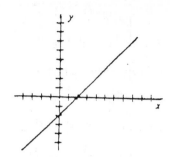

Figure 4

4) $x = t^3 + 1 \Rightarrow t^3 = x - 1$; $y = t^3 - 1 = x - 2$; As t varies from -2 to 2, (x, y) varies from $(-7, -9)$ to $(9, 7)$. See Figure 4.

7) $x = e^t \Rightarrow t = \ln x$; $y = e^{-2t} = e^{-2\ln x} = \left(e^{\ln x}\right)^{-2} = x^{-2} = \frac{1}{x^2}$; As t varies from $-\infty$ to ∞, x varies from 0 to ∞, excluding 0. See Figure 7.

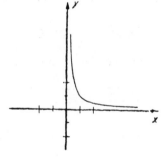

Figure 7

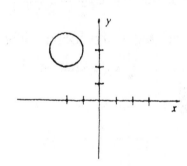

Figure 10

10) $x = \cos t - 2$ and $y = \sin t + 3 \Rightarrow x + 2 = \cos t$ and $y - 3 = \sin t$ $\Rightarrow (x + 2)^2 + (y - 3)^2 = \cos^2 t + \sin^2 t = 1$; As t varies from 0 to 2Π, (x, y) traces the circle from $(-1, 3)$ in a clockwise direction back to $(-1, 3)$. See Figure 10.

13) $x = t^2 \Rightarrow t = \sqrt{x}$ { since $t > 0$ }; $y = 2 \ln t = 2 \ln x^{1/2} = 2 \cdot \frac{1}{2} \ln x = \ln x$; As t varies from 0 to ∞, so does x, and y varies from $-\infty$ to ∞. See Figure 13.

Exercises 10.8

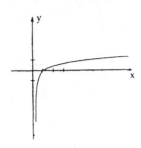

Figure 13

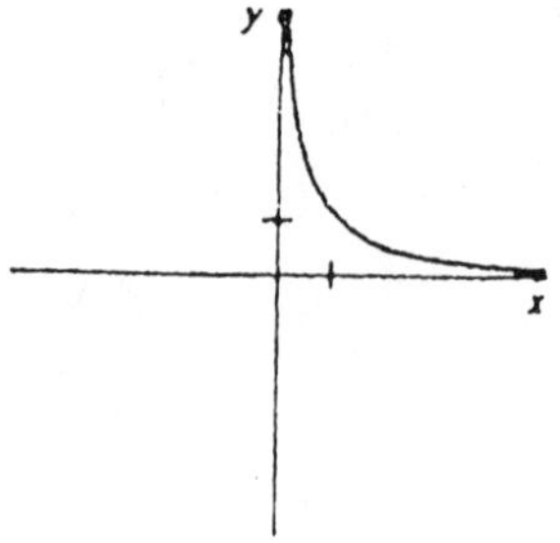

Figure 16

16) $x = e^t \Rightarrow t = \ln x;\ y = e^{-t} = e^{-\ln x} = \left(e^{\ln x}\right)^{-1} = x^{-1} = \frac{1}{x};$

As t varies from $-\infty$ to ∞, (x, y) varies from the positive x-axis to the positive y-axis. See Figure 16.

19) $x = t$ and $y = \sqrt{t^2 - 2t + 1} \Rightarrow y = \sqrt{x^2 - 2x + 1} = \sqrt{(x - 1)^2} = $
   $|x - 1|;$ As t varies from 0 to 4, (x, y) traces $y = |x - 1|$ from (0, 1) to (4, 3).See Figure 19.

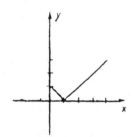

Figure 19

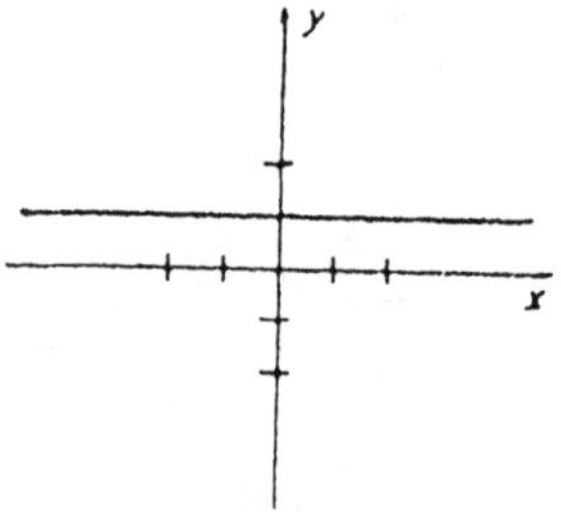

Figure 22

22) As t varies from $-\Pi/2$ to $\Pi/2$, x varies from $-\infty$ to ∞. y is always 1 so we have the graph $y = 1$. See Figure 22.

25) Solving the first equation for t we have $t = \frac{x - x_1}{x_2 - x_1}$. Substituting this expression into the second equation yields

$$y = (y_2 - y_1)\left(\frac{x - x_1}{x_2 - x_1}\right) + y_1 \Rightarrow y - y_1 = \left(\frac{y_2 - y_1}{x_2 - x_1}\right)(x - x_1) \text{ which is}$$

the Point-Slope Formula for the equation of a line through P_1 and P_2.

Other possibilities are :

$$x = (x_2 - x_1) \sqrt[3]{t} + x_1, \quad y = (y_2 - y_1) \sqrt[3]{t} + y_1$$
$$x = (x_2 - x_1) t^3 + x_1, \quad y = (y_2 - y_1) t^3 + y_1$$
$$x = (x_2 - x_1) t^5 + x_1, \quad y = (y_2 - y_1) t^5 + y_1$$

In general, $x = (x_2 - x_1) t^n + x_1$, $y = (y_2 - y_1) t^n + y_1$ are

parametric equations for l if n is any odd positive integer.

28) (appears on the

next page)

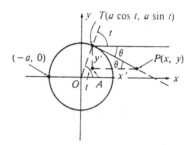

Figure 31

31) See Figure 31. Since the circle has radius a, the coordinates of the point of tangency are ($a \cos t$, $a \sin t$). Now if P is a typical point on the unraveling string, then the position of P is $x = a \cos t + x'$ and $y = a \sin t - y'$; that is, P is x' units to the right of the point of tangency and y' units below the point of tangency. We now seek to determine x' and y' in terms of t. $\overline{TP}$ has length ta, the arc length on the circle. Since the tangent line TP is perpendicular to the radius OT, $\theta + t = \Pi/2$ or equivalently, $\theta = \Pi/2 - t$. Hence, in the right triangle TAP, $\cos \theta = x'/ta \Rightarrow x' = at \cos (\Pi/2 - t) =$ at $\sin t$ and $\sin \theta = y'/ta \Rightarrow y' = at \sin (\Pi/2 - t) = at \cos t$. Thus, $x = a \cos t + at \sin t = a (\cos t + t \sin t)$ and $y = a \sin t - at \cos t = a (\sin t - t \cos t)$.

Exercises 10.8

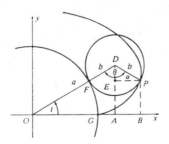

Figure 28

28) See Figure 28. Let $\theta = \angle FDP$ and $\alpha = \angle ADP$. Then $\angle ODA = \theta - \alpha$ $= (\Pi/2) - t$. Thus, $\alpha = \theta - [\,(\Pi/2) - t\,] = \theta + t - (\Pi/2)$. Arcs GF and PF are equal in length since each is the distance rolled. Thus $at = b\theta$ or $\theta = (a/b)\,t$ and $\alpha = (a/b)\,t + t - (\Pi/2) =$ $[\,(a + b)/b\,]\,t - (\Pi/2)$. Note that $\cos \alpha = \sin(\Pi/2 - \alpha) =$

$$\sin\left(\frac{\Pi}{2} - \left(\left(\frac{a+b}{b}\right)t - \frac{\Pi}{2}\right)\right) = \sin\left(\Pi - \left(\frac{a+b}{b}\right)t\right) = \sin\left(\frac{a+b}{b}t\right) \text{ and}$$

$$\sin \alpha = \cos\left(\frac{\Pi}{2} - \alpha\right) = \cos\left(\frac{\Pi}{2} - \left(\left(\frac{a+b}{b}\right)t - \frac{\Pi}{2}\right)\right) =$$

$$\cos\left(\Pi - \left(\frac{a+b}{b}\right)t\right) = -\cos\left(\frac{a+b}{b}\right)t. \quad \{\text{For the above use the}$$

difference identities for the sine and cosine.} For the location of the points as illustrated, the coordinates of P are :

$x = d\,(O, A) + d\,(A, B) = d\,(O, A) + d\,(E, P) = (a + b)\cos t + b\sin \alpha$

$\quad = (a + b)\cos t - b\cos[(a + b)/b]\,t$

$y = d\,(B, P) = d\,(A, D) - d\,(D, E) = (a + b)\sin t - b\cos \alpha$

$\quad = (a + b)\sin t - b\sin[(a + b)/b]\,t$

1) $y^2 = 64x \Rightarrow (y - 0)^2 = 64(x - 0)$; $4p = 64 \Rightarrow p = 16$;

V $(0, 0)$; F $(16, 0)$; $\ell : x = -16$; See Figure 1.

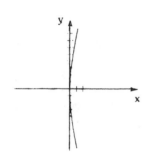 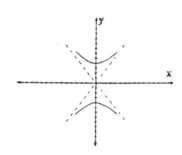

Figure 1   Figure 4

4) $9y^2 = 144 + 16x^2 \Rightarrow \dfrac{y^2}{16} - \dfrac{x^2}{9} = 1$; $c^2 = 16 + 9 \Rightarrow c = \pm 5$;

V $(0, \pm 4)$; W $(\pm 3, 0)$; F $(0, \pm 5)$; $y = \pm \frac{4}{3}x$; See Figure 4.

7) $25y = 100 - x^2 \Rightarrow (x - 0)^2 = -25(y - 4)$; $4p = -25 \Rightarrow p = -\frac{25}{4}$;

V $(0, 4)$; F $(0, -\frac{9}{4})$; $\ell : y = \frac{41}{4}$; See Figure 7.

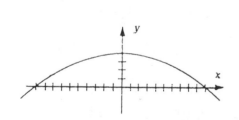 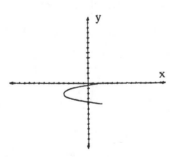

Figure 7   Figure 10

10) $x = 2y^2 + 8y + 3 \Rightarrow x = 2(y^2 + 4y + 4) + 3 - 8 \Rightarrow$

$(y + 2)^2 = \frac{1}{2}(x + 5)$; $4p = \frac{1}{2} \Rightarrow p = \frac{1}{8}$; V $(-5, -2)$;

F $(-\frac{39}{8}, -2)$; $\ell : x = -\frac{41}{8}$; See Figure 10.

13) F $(0, -10)$ and $\ell : y = 10 \Rightarrow p = -10$ and V $(0, 0)$; An equation is

$(x - 0)^2 = -40(y - 0)$ or $x^2 = -40y$.

16) F ($\pm$ 10, 0) and V ($\pm$ 5, 0) $\Rightarrow$ $b^2 = 10^2 - 5^2 = 75$; An equation is

$\dfrac{x^2}{25} - \dfrac{y^2}{75} = 1$ or $3x^2 - y^2 = 75$.

19) $4x^2 + 9y^2 + 24x - 36y + 36 = 0$ $\Rightarrow$

$4(x^2 + 6x + 9) + 9(y^2 - 4y + 4) = -36 + 36 + 36 = 36$ $\Rightarrow$

$\dfrac{(x+3)^2}{9} + \dfrac{(y-2)^2}{4} = 1$; $c^2 = 9 - 4$ $\Rightarrow$ $c = \pm\sqrt{5}$; C (-3, 2);

V (-3 $\pm$ 3, 2); M (-3, 2 $\pm$ 2); F (-3 $\pm$ $\sqrt{5}$, 2); See Figure 19.

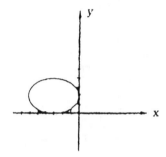

Figure 19

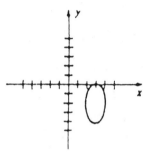

Figure 22

22) $4x^2 + y^2 - 24x + 4y + 36 = 0$ $\Rightarrow$

$4(x^2 - 6x + 9) + (y^2 + 4y + 4) = -36 + 36 + 4 = 4$ $\Rightarrow$

$\dfrac{(x-3)^2}{1} + \dfrac{(y+2)^2}{4} = 1$; $c^2 = 4 - 1$ $\Rightarrow$ $c = \pm\sqrt{3}$; C (3, -2);

V (3, -2 $\pm$ 2); M (3 $\pm$ 1, -2); F (3, -2 $\pm$ $\sqrt{3}$); See Figure 22.

25) $r = -4\sin\theta$ $\Rightarrow$ $r^2 = -4r\sin\theta$ $\Rightarrow$ $x^2 + y^2 + 4y = 0$ $\Rightarrow$

$x^2 + (y+2)^2 = 4$; {circle} See Figure 25.

28) $0 = 9\sin 2\theta$ $\Rightarrow$ $\sin 2\theta = 0$ $\Rightarrow$ $2\theta = \Pi n$ $\Rightarrow$ $\theta = (\Pi/2)n$

	Range of θ		Range of r	
1)	0 $\rightarrow$	$\Pi/4$	0 $\rightarrow \pm$ 3	
2)	$\Pi/4$ $\rightarrow$	$\Pi/2$	$\pm$ 3 $\rightarrow$ 0	
3)	$\Pi/2$ $\rightarrow$	$3\Pi/4$	undefined	
4)	$3\Pi/4$ $\rightarrow$	Π	undefined	See Figure 28.

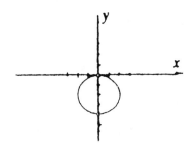

Figure 25

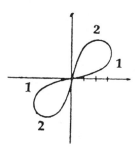

Figure 28

31) $r = 6 - r\cos\theta \implies r = 6 - x$

$\implies r^2 = 36 - 12x + x^2 \implies$

$y^2 = 36 - 12x$ or $x =$

$-\frac{1}{12} y^2 + 3$; {parabola}

See Figure 31.

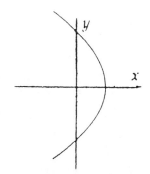

Figure 31

34) $x^2 + y^2 - 3x + 4y = 0 \implies r^2 - 3r\cos\theta + 4r\sin\theta = 0 \implies$

$r - 3\cos\theta + 4\sin\theta = 0$ or $r = 3\cos\theta - 4\sin\theta$

37) $r^2 = \tan\theta \implies x^2 + y^2 = \frac{y}{x} \implies x^3 + xy^2 = y$

40) $\theta = \sqrt{3} \implies \tan^{-1}\left(\frac{y}{x}\right) = \sqrt{3} \implies \frac{y}{x} = \tan\sqrt{3} \implies y = (\tan\sqrt{3})\,x$;

Note that $\tan\sqrt{3} \simeq -6.15$; This is a line through the origin making an angle of approximately 99.24^0 with the positive x-axis. The line is not $y = (\Pi/3)\,x$.

43) $x = \sqrt{t} \implies t = x^2$ and $y = 2^{-x^2}$; The graph is a bell-shaped curve with a maximum point at $t = 0$ or $(0, 1)$. As t increases, x increases, and y gets close to 0. See Figure 43.

Exercises 10.9

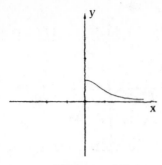

Figure 43

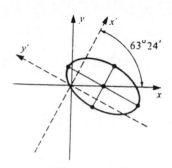

Figure 46

See Exercises 10.5 for the general solution outline for Exercise 46.

46) (a) $\cot 2\phi = -\frac{3}{4}$; $\phi \approx 63.43°$; See Figure 46.

(b) $\cos 2\phi = -\frac{3}{5}$; $\sin \phi = 2/\sqrt{5}$; $\cos \phi = 1/\sqrt{5}$

(c) $\begin{cases} x = (1/\sqrt{5}) x' - (2/\sqrt{5}) y' = (1/\sqrt{5}) (x' - 2y') \\ y = (2/\sqrt{5}) x' + (1/\sqrt{5}) y' = (1/\sqrt{5}) (2x' + y') \end{cases}$

(d) $\frac{100}{5} (x')^2 + \frac{25}{5} (y')^2 - 40x' + 20y' = 0 \Rightarrow$

$4 (x')^2 - 8x' + (y')^2 + 4y' = 0 \Rightarrow$

$4 ((x')^2 - 2x' + 1) + ((y')^2 + 4y' + 4) = 4 + 4 = 8 \Rightarrow$

$\frac{(x' - 1)^2}{2} + \frac{(y' + 2)^2}{8} = 1$; ellipse

(e) $C'(1, -2)$; $V'(1, -2 \pm 2 \sqrt{2})$; $M'(1 \pm \sqrt{2}, -2)$

(f) $C (\sqrt{5}, 0)$; $V ((5 \mp 4 \sqrt{2})/\sqrt{5}, (\pm 2 \sqrt{2})/\sqrt{5})$;

$M ((5 \pm \sqrt{2})/\sqrt{5}, (\pm 2 \sqrt{2})/\sqrt{5})$

Exercises A.1

For the solutions in this appendix, it is assumed that the reader is familiar with all the properties of the logarithm. The = sign is used for table values even though the case could be made that $\approx$ should be used. The values are not those that you would get using a calculator, and sufficient work is shown as to how the answers were arrived at.

1) $\log 3.47 \times 10^k = 0.5403 + k$; $k = 2, -3, 0 \Rightarrow$

$$2.5403,\ \ 7.5403 - 10,\ \ 0.5403$$

4) $\log 2.08 \times 10^k = 0.3181 + k$; $k = 2, 0, 4 \Rightarrow$

$$2.3181,\ \ 0.3181,\ \ 4.3181$$

7) $\log (44.9)^n = n \log 44.9 = n\,(1.6522)$; $n = 2, \frac{1}{2}, -2 \Rightarrow$

$$3.3044,\ \ 0.8261,\ \ 6.6956 - 10$$

10) $\log (0.017)^{10} = 10 \log 0.017 = 10\,(8.2304 - 10) = 82.304 - 100$

or $2.304 - 20$; $\log 10^{0.017} = 0.017$; $\log 10^{1.43} = 1.43$

13) $\log \dfrac{(47.4)^3}{(29.5)^2} = 3 \log 47.4 - 2 \log 29.5$

$$= 3\,(1.6758) - 2\,(1.4698) = 2.0878$$

16) $\log \dfrac{(0.0048)^{10}}{\sqrt{0.29}} = 10 \log 0.0048 - \frac{1}{2} \log 0.29$

$$= 10\,(7.6812 - 10) - \tfrac{1}{2}\,(9.4624 - 10)$$

$$= 72.0808 - 95 \text{ or } 0.0808 - 23$$

19) $\log x = 0.9469 \Rightarrow x = 8.85 \times 10^0 = 8.85$

22) $\log x = 6.7300 - 10 \Rightarrow x = 5.37 \times 10^{-4} = 0.000537$

25) $\log x = 8.8306 - 10 \Rightarrow x = 6.77 \times 10^{-2} = 0.0677$

28) $\log x = 3.0043 \Rightarrow x = 1.01 \times 10^3 = 1010$

31) Log $2.54 = 0.4048$ and $\log 2.55 = 0.4065$. The difference between 2.54 and 2.55 is 0.01. It is helpful to think of this difference as 10 units. We want eight-tenths of this difference. The difference between 0.4048 and 0.4065 is 0.0017. The log of 25.48 will then be 1.4048 plus $\frac{8}{10}$ of 0.0017. We will use the following shorter version to compute logarithms involving interpolation.

$\log 25.48 = 1.4048 + (.8)\,(.0017) = 1.4062$

34) $\log 0.3817 = 9.5809 - 10 + (.7)\,(.0012) = 9.5817 - 10$

Exercises A.1

37) $\log 123{,}400 = 5.0899 + (.4)(.0035) = 5.0913$

40) $\log 1.203 = 0.0792 + (.3)(.0036) = 0.8030$

43) $\log 0.9462 = 9.9759 - 10 + (.2)(.0004) = 9.9760 - 10$

46) $\log 0.001428 = 7.1523 - 10 + (.8)(.0030) = 7.1547 - 10$

49) $\log 3.003 = 0.4771 + (.3)(.0015) = 0.4776$

52) $\log x = 3.7455 \Rightarrow x = 5560 + \frac{4}{8}(10) = 5565$

55) $\log x = 9.1664 - 10 \Rightarrow x = 0.146 + \frac{20}{29}(.001) = 0.1467$

58) $\log x = 5.9306 - 9 \Rightarrow x = 0.000852 + \frac{2}{5}(.000001) = 0.0008524$

61) $\log x = 0.1358 \Rightarrow x = 1.36 + \frac{22}{32}(.01) = 1.367$

64) $\log x = 2.4979 - 5 \Rightarrow x = 0.00314 + \frac{10}{14}(.00001) = 0.003147$

67) $\log x = -2.8712 = 7.1288 - 10 \Rightarrow x = 0.00134 + \frac{17}{32}(.00001)$
$$= 0.001345$$

70) $\log x = -3.1426 = 6.8574 - 10 \Rightarrow x = 0.000720 + \frac{1}{6}(.000001)$
$$= 0.0007202$$

73) $\tan 3 = -\tan(\Pi - 3) = -\tan(0.1416) = -(.1405 + \frac{20}{29}(.0030))$
$$= -0.1426$$

76) $\csc 1.54 = 1.001 - \frac{12}{29}(.001) = 1.0006$

79) $\cot 62^0 27' = 0.5243 - \frac{7}{10}(.0037) = 0.5217$

82) $\sec 9^0 12' = 1.013 + \frac{2}{10}(.000) = 1.013$

85) $\tan t = 4.501 \Rightarrow t = 1.3497 + \frac{52}{62}(.0029) = 1.3521$

88) $\cot t = 1.165 \Rightarrow t = 0.7079 + \frac{5}{7}(.0029) = 0.7094$

91) $\tan \theta = 0.5042 \Rightarrow \theta = 26^0 40' + \frac{20}{37}(10') = 26^0 45'$;
$180^0 + 26^0 45' = 206^0 45'$

94) $\csc \theta = 1.219 \Rightarrow \theta = 55^0 00' + \frac{2}{3}(10') = 55^0 07'$;
$180^0 - 55^0 07' = 124^0 53'$